Hans Palmgren

The V-belt Handbook

Studentlitteratur   Chartwell-Bratt

This book was first published in Swedish in 1985

The title of the original edition: Kilremshandboken

Printed in Sweden
Studentlitteratur
Lund 1986
ISBN/Studentlitteratur 91-44-22961-5
ISBN/Chartwell-Bratt 0-86238-111-8

Contents

4

6

Preface

This book is the English, slightly modified version of the Swedish original which was published 1985. Its main subject is V-belts with brief outlooks on other types of belts. The book is written for those directly engaged in technology, engineering, service and development involving belt drives or others who want to widen and deepen their knowledge within this area. It is also hoped that the book may serve as a textbook for teachers as well as for students when studying machine or transmission design. The book is meant to fill the information gap within this specialized area where – as far as known – no similar treatise is at hand.

The book is directed towards what is technically and practically useful. The mathematical treatment is limited to what is normally used in engineering design calculations. More exact and usually more elaborate methods of calculation can be found in the references cited in Appendix B. To satisfy readers who prefer a discussion without mathematical treatment or equations the first ten chapters are totally free of mathematics. This is obviously in contrast to most textbooks where as a rule the theory and mathematical treatment come first after which the practical applications are discussed.

All quantities are expressed in SI units in their basic form if not otherwise indicated. Appendix A contains a list of quantities used, their symbols and corresponding SI units. The symbols used as well as the vocabulary in general is, as far as possible, in agreement with that used in international (ISO) standards. A belt drive vocabulary in Appendix H will facilitate the search for connections between V-belt terms in four languages.

The importance of standardization in this area, especially international standardization, is stressed throughout the book and especially in chapter 10 and Appendix C.

Correct design calculations of belt drives require knowledge of a great number of data for the belt types to be used. Tables and charts given in the text as well as in Appendix D, E, F and G are only to be considered as examples of such data and to illustrate the sample belt design calculation given in chapter 18.

A detailed subject index is given in Appendix I as an identification aid to a given problem area. Cross references are frequently used in the text for the same purpose.

This book has only been possible by valuable support and help from many sides for which special thanks are due to: The original idea for the book was generated within the business area Industrial Supply, Transmission Division, at Trelleborg AB who thereafter has generously financed the project and supported it in many ways. Many colleagues at Trelleborg AB have contributed with their specialist knowledge, above all Ulf Boman, Karl Axel Fridh, Lars-Peter Holmlund, Bo Larsson, Gunnar Lindén and Hans Tjörnvik. The extensive work with the manuscript, involving textual revision, typing and checking, has skilfully and carefully been handled by Anne Wängberg. Elisabeth Håkansson has assisted in procuring the illustrations, Rolf Lygård (at Teknisk Information) has drawn the sketches and graphs and Birgit Norén has done some of the typing. My English text has been skilfully revised by Gaston Demaret. Finally I want to thank my wife Maj-Britt who together with me has devoted many evenings at home to reading aloud technical articles covering the very fascinating technique area represented by V-belts.

Trelleborg in February 1986

Hans Palmgren

1. Introduction

1.1 Historical

It is often said that the main difference between man and animal is man's ability to utilize tools and implements. Already pre-historic man learned to use tools and to multiply his force by applying the principle of leverage. The transmission of power to a rotating shaft, e g for the grinding of grain or the pumping of water, gradually became the main field of power transmission. Already during the Roman era we find the predecessors of present-day gear drives. Power transmissions utilizing belts were also known in early times, figure 1-1. During the industrial revolution in the early 19th century the need for mechanical power transmissions increased dramatically. The problem was usually to transmit power from a centrally placed power source, e g a water

Figure 1-1. Early machinery with belt drives (Ravelli, 1558)

wheel or a steam engine, to a number of dependent machines. Using rotating shafts, connected to the centrally placed power source, power was transmitted to individual machines by belt drives. Flat belts, mainly made from leather, were used. Around the turn of the century flat textile belts were developed, with or without impregnation and cover, usually of rubber, to improve friction and durability.

It was then found that the use of ropes or wires, running in grooves in pulleys, could increase the transmitted power considerably with the help of the wedge effect. V-belts, with a trapezoidal section specially developed to fit the grooves in the pulleys, were developed during the 1920's, primarily as fan belts in cars. These quickly spread to the industrial market replacing the flat belts. As the V-belts were taking over the market the old system of a central driving unit servicing several driven machines was being abandoned and replaced by separate driving units (e g with electrical motors) for each driven machine. Subsequent improvements in the materials, construction and manufacturing methods of V-belts have perpetuated a continuous improvement of power ratings, cf figure 6-1.

1.2 Some principles of mechanical power transmission

For mechanical transmission various principles and devices have been developed, often well described in pertinent textbooks and handbooks. A condensed tabular summary with comparative, characteristic data can be found in figure 1-2. The table points to some data which has to be considered in the selection of a suitable mechanical power transmission. The most important factors to be considered are usually the power or torque to be transmitted and the rotational speeds of the driving and driven shafts. In figure 1-3 typical ranges of these factors for some applications are shown and in figure 1-4 corresponding performance characteristics for important types of mechanical power transmission devices are given. Other important factors to be considered are, for example, the need for a constant or variable speed ratio, precision in the transmission of speed and the restrictions on shaft centre distance. Noise level, reliability, maintenance and space requirements may also be of great importance. When all technical requirements have been met the final choice is then made taking into account the economical considerations, which should include acquisition costs, maintenance and service costs as well as costs for energy losses.

Type Typical data	Gear drive	Friction wheel drive	Chain drive	Hydraulic transmission	V-belt drive	Flat belt drive	Synchronous belt drive
Power transmission by interlocking (I) or by friction (F)	I	F	I	–	F	F	I
Max power rating P (kW) Normal	10 000	20	500	1 000	1 000	1 000	500
Extreme	100 000	200	1 000	10 000	2 000	5 000	1 000
Max torque M (Nm)	10^8	10^3	10^6	10^8	10^4	10^4	10^4
Max peripheral speed v (m/s) Normal	50	25	15	–	30	50	60
Extreme	100	50	30	–	50	120	100
Max speed ratio c_p Normal	15:1	9:1	6:1	100:1	10:1	15:1	15:1
Extreme	50:1	15:1	15:1	∞:1	15:1	25:1	50:1
Accuracy of speed ratio c_p	high	moderate	high	moderate	moderate	moderate	high
Efficiency η (%) at full load	94–98	86–94	94–98	84–92	92–97	92–97	93–98
Operating temperature (°C) Normal	−30/+150	−30/+150	−30/+150	−20/+100	−40/+80	−40/+80	−40/+80
Extreme	−30/+250	−30/+250	−30/+250	−30/+150	−50/+120	−50/+120	−50/+120
Noise level	high	high	high	low	low	low	moderate
Lubrication required	yes	occasionally	yes	–	no	no	no

Figure 1-2. Comparative, approximate data for various power transmission devices

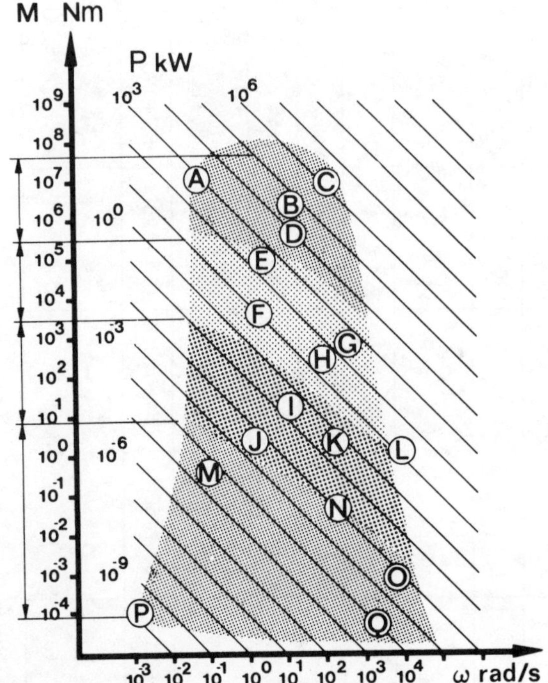

Figure 1-3. Typical requirements for torque M (Nm) and angular velocity ω (rad/s) in various types of equipment. In the diagram lines are drawn to indicate different levels of power P (kW)

A. Rotary cement kiln
B. Gas turbine in container ship
C. Electricity power generator
D. Diesel engine in ferry
E. Caterpillar
F. Windmill
G. Refrigeration compressor
H. Volvo 340 car
I. Washing machine
J. Windshield wiper motor
K. Hand power tool
L. Truck turbo charger
M. Timer clock
N. Electric razor
O. Gyroscope
P. Recording cylinder
Q. Dentist's drilling machine

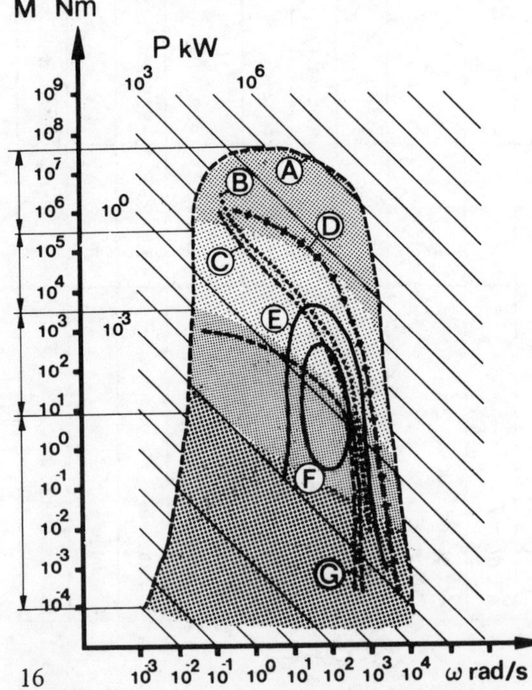

Figure 1-4. Use of various power transmission devices with respect to magnitude of torque M (Nm) and angular velocity ω (rad/s). In the diagram lines have been drawn to indicate various levels of power P (kW)

A. Cylindrical gear drive
B. Chain drive
C. Worm or hypoid gear drive
D. Conical gear drive
E. Belt drive (V-belts, flat belts or synchronous belts)
F. Adjustable-speed drive with variator belts
G. Helical gear drive

1.3 Advantages of belt drives

Compared to other mechanical power transmission devices belt drives offer the following advantages:

- Low maintenance costs (lubrication not required)
- Optional choice of speed ratio and shaft centre distance (within certain limits)
- Simple installation
- Perfect alignment of shafts usually not required
- Low space requirements by utilization of the flexibility offered in mounting
- High degree of reliability
- The elasticity of the belts dampens shocks and protects the equipment against sudden overloads
- Belts and pulleys are well standardized and easily available as stock items and spare parts
- Low acquisition cost and good economy

2. Some fundamental principles of belt drives

2.1 Introduction

In this chapter some fundamental aspects of belt drives will be briefly discussed. Reference to the more detailed discussion in following chapters will be made where applicable.

A belt drive is composed of one or several belts running over two or more pulleys or sheaves mounted on driving and driven shafts respectively. Sometimes idler pulleys are included to keep the belt under tension and increase the arc of contact between belt and pulleys, cf section 14.5. The simplest case, including two pulleys, is shown in figure 2-1. Various types of belts are discussed in chapter 3, materials used in chapter 6 and manufacturing methods in chapter 7. Various types of pulleys or sheaves are described in chapter 5.

2.2 Geometry

Equations for calculation of belt geometry are given in chapter 11. Important factors for the simple case are indicated in figure 2-1. The small pulley is usually driving (the driver) and the large one driven.

The diameter ratio of the two pulleys determines the speed ratio of the belt drive, i e the ratio of the rotational speeds of the driving and driven shaft respectively. A stepwise variable speed ratio can be obtained by placing a set of pulleys of different diameters, together on the same shaft and then moving the belt between them. These are termed stepped pulleys, cf figure 5-6. Continuously variable speed ratios can be obtained by the use of so-called variator drives in which the pulley is split, in its own plane, in two axially movable halves, figure 2-2. With the aid of mechanical devices the two halves can be axially moved changing the width of the wedge-shaped gap between them. When the width of the gap is increased the belt moves radially down-

18

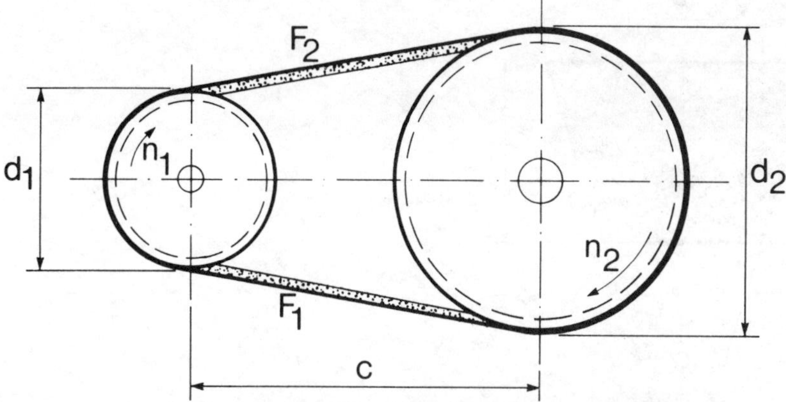

Figure 2-1. Simple belt drive with two pulleys in the same plane. The driving pulley (usually the small one) with diameter d_1 rotates with n_1 rpm and the driven pulley (usually the large one) with diameter d_2 rotates with n_2 rpm. The tight (pulling) side has belt tension F_1 and the slack side belt tension F_2. Shaft centre distance is denoted by c.

wards, deeper in the groove, and will thus operate at a smaller diameter. Inversely the belt will move radially upwards in the groove and operate at a larger diameter when the two pulley halves are moved closer together. In a V-belt variator drive either only one, usually the driven one, or both pulleys are split and axially adjustable. When only one pulley is split the variator allows a speed ratio change of up to 3:1, and when both pulleys are split of up to 9:1. If only one of the pulleys in a variator drive is split the shaft centre distance must be adjustable when speed ratio is varied. Sometimes a non-adjustable shaft centre distance is used and changes of axial force are used to adjust belt tension on the whole maintaining speed ratio, cf section 14.8. If both pulleys are split speed ratio can be adjusted without changes in shaft centre distance.

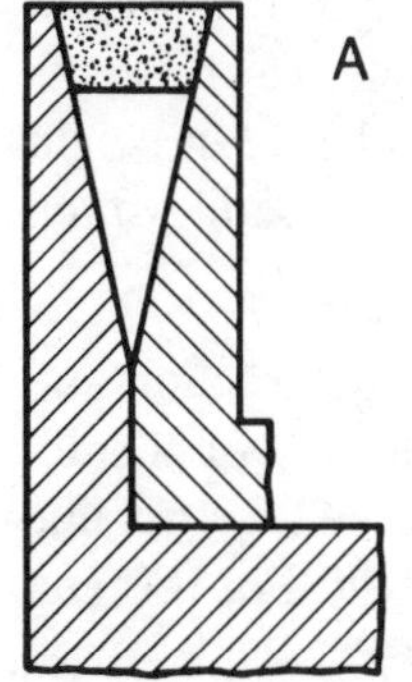
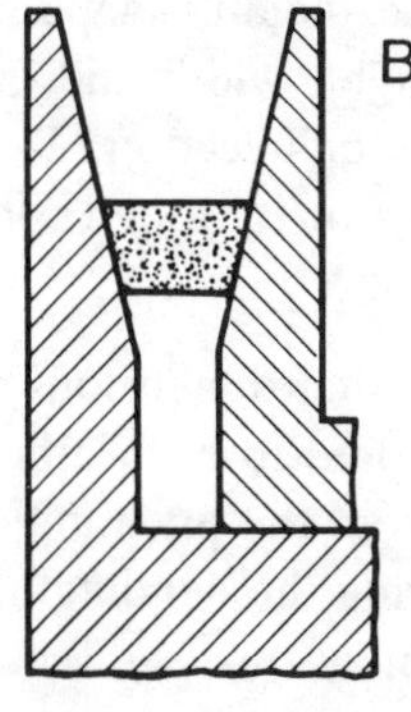

Figure 2-2. Principle of adjustable-speed drive with variator belt
A. Pulley halves close together, large pitch diameter
B. Pulley halves apart, small pitch diameter

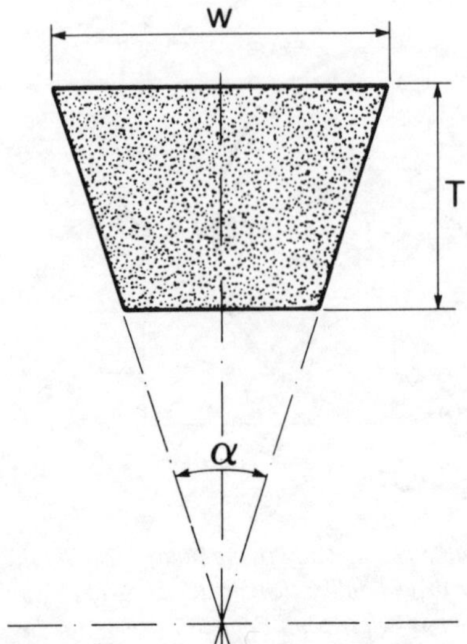

Figure 2-3. The V-belt section is trapezoidal, characterized by width w, height T and belt angle α

The belt length can be estimated or calculated from the shaft centre distance and the diameters of the pulleys, cf figure 8-1. With selection of standard belt lengths and standard pulley diameters the shaft centre distance must be adjusted to reach agreement with the geometrical calculations. If the belt drive contains several, parallel running belts (multiple belt drive) the difference in length between individual belts must be small. In this way an even force distribution between the belts is maintained and thus the optimal power rating and durability of the belt drive reached. Belts of standardized length were earlier grouped into classes with narrow tolerances, from which a matched set of belts could be picked out for a multiple belt drive. Because of improvements in manufacturing technology belt length tolerances can nowadays be kept so small that all belts of a standardized length can be used in matched sets for multiple belt drives.

The cross-section of a V-belt is trapezoidal in shape, figure 2-3. The section can thus be characterized by its height, width and belt angle, i e the angle between the sloping sidewalls of the belt section. The ratio between height and width (pitch width) of the belt section is termed relative height and is often used to characterize a certain type of V-belt.

A satisfactory V-belt drive requires that dimensions of the belt section and grooves in the pulleys are well adapted to each other. One way of characterizing the dimensions is to start from the pitch line of the belt, i e the line which doesn't change its length during the bending of the belt. For this and other methods of characterizing dimensions, see section 11.2.

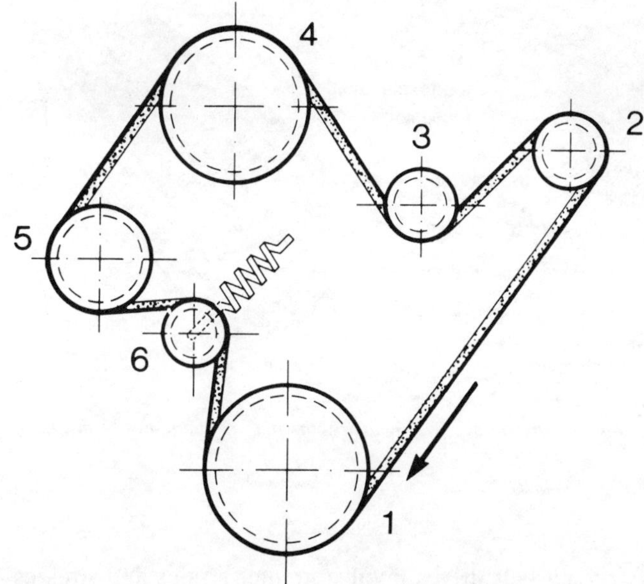

Figure 2-4. Example of belt drive with several pulleys operating on the inside as well as on the outside of the belt, so-called serpentine drive. The example illustrates an automotive belt drive where the various pulleys are connected to: 1. crank-shaft, 2. power steering, 3. compressor, 4. fan and water pump and 5. generator. In addition an idler pulley 6 is used.

The geometry will be more complicated in belt drives utilizing more than two pulleys, figure 2-4. Practical applications of this type are, fore instance, found in automotive and agricultural belt drives.

Belt drives normally have parallel shafts and pulleys in the same plane. In some applications, especially in agricultural drives, shafts are not parallel but crossing each other at an angle, so-called twisted belt drives, figure 2-5. If shafts cross each other at an angle of 90° the arrangement is often termed quarter-turn drive. Crossed belt drives have shafts that cross each other at an angle of 180°, which reverses the direction of rotation between driving and driven shafts. The shafts are still parallel but the strands of the belt cross each

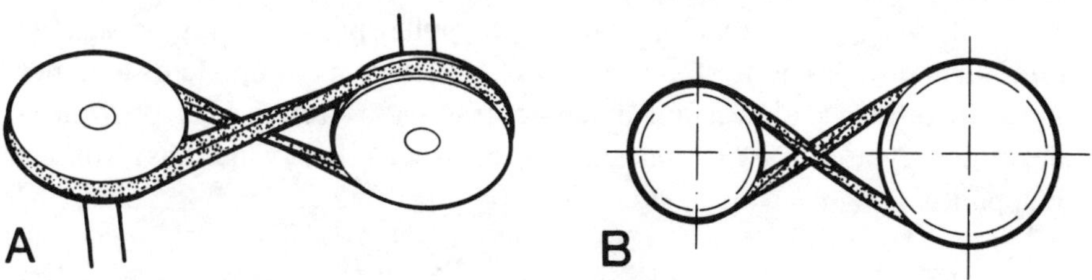

Figure 2-5. Quarter turn (A) and crossed (B) belt drives

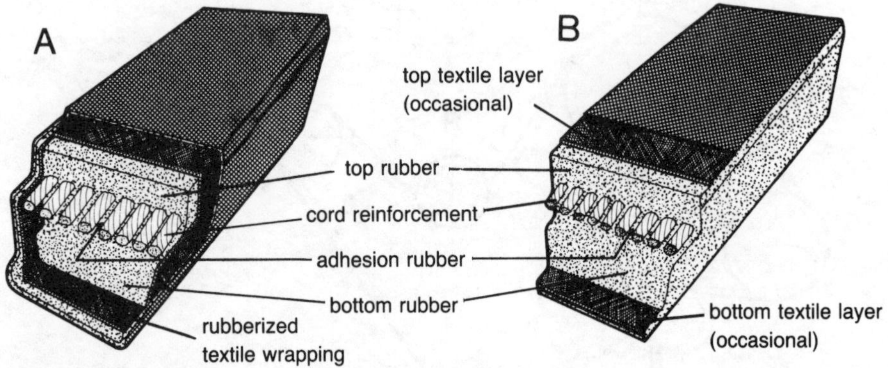

Figure 2-6. Components in a V-belt
A. With textile wrapping, longer belts (above 2,5 m in length) may contain several layers of
*   reinforcement*
B. Without textile wrapping (cut V-belt)

other, cf figure 2-5. Crossed belt drives involve greater strains and stresses on the belts. There is a risk for tilting, running off the grooves or wear by the strands chafing each other. They are thus only used in applications where some sacrifice in power rating and/or useful life is accepted, cf section 19.4.

2.3 Belt construction

The fundamental construction of a V-belt is shown in figure 2-6. The load-carrying component of the V-belt is the reinforcement, usually textile cord. The reinforcement is normally placed just under the upper surface of the belt section. The pitch zone of the belt will also be found here. The cords are embedded in soft rubber, especially compounded to confer good adhesion. The section below the reinforcement is filled up with bottom rubber and the section above the reinforcement with top rubber. The belt section can be surfaced with a textile wrapping impregnated with rubber, to increase the friction between the belt and the sides of the pulley groove, to protect against abrasion and wear and to stabilize the belt and keep its components together under the influence of fatiguing dynamic strains and stresses when operational. Another type of construction is the so-called cut V-belt, which has no textile wrapping.

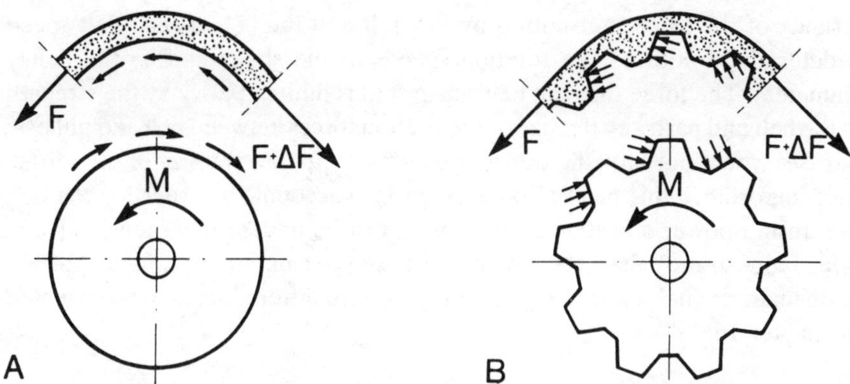

Figure 2-7. Power transmission utilizing friction (A) or positive engagement with mechanical interlocking (B). To transmit torque M a belt tension difference $\triangle F$ between tight and slack side is necessary, which for V-belts is counterbalanced by friction forces and for synchronous belts by compressive forces on the surfaces of the cogs

2.4 Power transmission

The transmission of power between belt and pulleys can either involve frictional forces or positive engagement and mechanical interlocking, for instance with the aid of cogs or teeth, figure 2-7. The frictional forces, and thus the possible power rating, will depend on the coefficient of friction, the normal force and the contact area between belt and pulleys. The size of the contact area will depend on factors such as the arc of contact angle θ, figure 2-1. Any increase in contact force between belt and pulleys will cause increased belt tension meaning increased loads on shafts and shaft bearings. One method of increasing frictional force without a corresponding increase of the belt tension is to utilize the wedge effect, as in a V-belt, figure 12-1. The wedge effect was originally applied to ropes or wires, mainly with circular sections in wedge profile pulleys, but is today the basis for the widespread use of V-belts. Belt angle shall be chosen to give adequate enlargement of the friction force but not so small that belts are locked in the grooves. It should be noted, cf section 5.2.1, that the belt angle diminishes when belts are bent over the pulleys, and this increases the risk of the belt locking. A mathematical analysis of the forces between belts and pulleys is presented in chapter 12.

The power transmission capability of a belt drive is governed by several factors, some restricted by system requirements and available equipment, while others are not restricted. The power transmitted can be written as the product of the force transmitted by the belt and the belt speed. Belt speed is determined both by the rotation speeds of the shafts and by the pulley diameters. The force that can be transmitted is limited partly by the strength of the belt and partly by the size of the friction forces between belt and pulleys. In cases where not only the transmitted power but also the size of the torque is of importance this must also be taken into account. In order for the belt to transmit power one strand of the belt must be under higher tension (tight side, tight strand) than the other (slack side, slack strand), figure 2-1. A mathematical analysis of the interaction of these different factors is presented in chapter 16.

2.5 Stresses and strains in belts

The following types of stresses and strains can be found in a belt:

- *Tensile stresses*, needed to give contact forces and transmission of power
- *Tensile stresses*, caused by centrifugal forces during the rotation of the pulleys
- *Dynamic tensile stresses*, caused by the changes in force level between tight side and slack side of the belt
- *Bending stresses*, caused by the bending of the belts over the pulleys
- *Compression or shear stresses*, caused by the transmission of forces between belts and pulleys

To the above should be added abrasive action, especially where the belts enter or leave the pulleys, and chemical deterioration caused by increase in working temperature (e g caused by heat generation during dynamic operation of the belt) or surrounding chemicals such as oil or aggressive components in the atmosphere such as oxygene or ozone. A mathematical analysis of stresses and strains in belts is presented in chapter 13. Other factors that influence the life of the belts are discussed in secton 4.2 and chapter 6.

2.6 Belt tension and loads on bearings

In order to reach the required frictional forces, in cases where power is transmitted by friction, and adequate level of contact forces between belts and pulleys will be required. The necessary contact force is obtained by using the belt under tension, i e exposed to tensile forces. The forces caused by the belt tension will not only affect the belt but also shafts and bearings. Too high belt tension (too large belt forces) increases stresses and strains on the components of the belt drive and will shorten their useful life. Too low belt tension (too small belt forces) will decrease power transmission capability and may also lead to slippage and consequent destruction of the belts. Choice of proper belt tension and various devices to keep the belts under tension are discussed in chapter 14.

2.7 Belt life

Belts are, when working, exposed to repeated stresses and strains, which means that their useful working life is determined by fatigue phenomena, discussed in chapter 15. Incorrect installation and abnormal loading conditions can lead to catastrophic wear of the belts resulting in a very short useful life, see section 9.6.

2.8 Efficiency

Soaring energy costs have increased the importance of high mechanical efficiency of a belt drive, requiring that as large a portion as possible of the power from the driving shaft is transmitted to the driven shaft. Furthermore, any mechanical energy loss is transformed into heat in the belts, leading to increased working temperature and shortened belt life. The efficiency of belts and belt drives is discussed in chapter 17.

2.9 Belt drive design calculations

Design calculations of belt drives start with given data and requirements, for instance power to be transmitted, rotational speed of driving and driven shafts, desired shaft centre distance, mechanical and chemical drive conditions, etc, and are expected to lead to recommendations of suitable types and number of belts, as well as types and diameters of pulleys. For the simple case including two pulleys operating in the same plane, figure 2-1, well-standardized design calculation methods have been in use for a long time, see chapter 18. For more complicated cases, e g involving more than two pulleys or introducing belt life as a variable, more advanced calculation methods have been developed, often to be performed with the aid of computers, see chapter 19.

3. Belt types

3.1 Introduction

In this chapter various types of belts are discussed. The description is concentrated on the most important ones. In addition a great number of varieties exist, which had to be neglected here due to space limitations. Characteristic data is presented in tables on typical sizes of a given type, usually taken from standards or manufacturers' catalogues. Data is often approximate and is only intended to give an indication of typical magnitude. When engineering calculations are to be made more exact data should be used taken from relevant standards or manufacturers' catalogues. The power ratings presented can only be reached by use of optimum belt speed, large enough pulley diameters and favourable drive conditions. For engineering calculations of power transmission capability in a given belt drive, see chapters 18 and 19.

3.2 Classical V-belts

The classical V-belt is the original V-belt construction, developed during the 1920's. It has a trapezoidal section and is usually enclosed in textile wrapping, figure 3-1. The relative height of the section is around 0,7 and the belt angle is chosen so that the V-belts will run properly in pulley grooves with an angle of 32–38°. The underside of classical V-belts is usually plane as the use of the wrapping makes cogging difficult in manufacture. Compared to many more modern types of belts the weight (mass) of classical V-belts is fairly large. This leads to high centrifugal forces and limits belt speed to roughly 25 m/s. Belt speed is further limited by the slow dissipation of the heat generated in the heavy belt section by the mechanical damping of bending stresses. Bending moment is also fairly high, which imposes restrictions on the minimum pulley diameter that can be used.

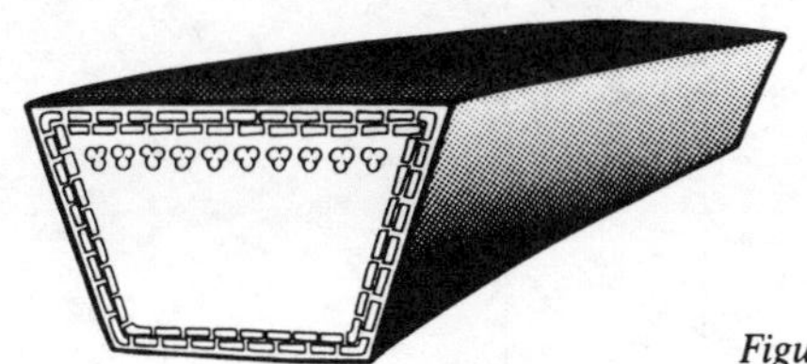

Figure 3-1. Principal design of classical V-belt

Typical data for standardized belt sections is presented in figure 3-2. For power transmission, in the medium range, use of multiple belt drives with A or B sections is preferred to single belt drives in C, D or E sections which are reserved for multiple belt drives in the high power range.

Classical V-belts were originally developed as fan belts in automobiles but are nowadays mainly used in industrial applications requiring good reliability and a long working life. With the introduction of improved types of V-belts their market share is now decreasing.

The low relative height of the classical V-belt section compared, for instance, to the narrow V-belt section makes these belts especially suitable in applications where they are bent over an idler operating on the outside of the belt. This is common in agricultural applications where the classical V-belt is the most widely used type. For agricultural applications an "H" is often used in front of the belt size designation, e g HA, HB, etc.

A special variant of the classical V-belts is the so-called single V-belt, mainly intended for the single V-belt drives of house-hold equipment with low power requirements, and standardized for instance in the American standard RMA IP-23. They are often termed FHP (Fractional Horse Power) V-belts. Standardized sections are designated 2L, 3L, 4L and 5L, which in order correspond to the classical V-belt sections Y, Z, A and B. These V-belts and their designations are mainly used in the USA while in Europe the corresponding classical V-belt sections are preferred. As the use of FHP V-belts usually involves low power requirements and intermittent drive conditions, their construction is mainly aimed to offer good flexibility and thereby allow small pulley diameters to be used rather than to give a long working life at high power ratings.

Profile designation (ISO 4184)	Y	Z	A	B	C	D	E
Profile dimensions, mm (DIN 2215)	6 / 4	10 / 6	13 / 8	17 / 11	22 / 14	32 / 20	40 / 25
Max basic power rating per belt, kW (SMS 2479-2485)	0,6	2,3	3,3	6,4	14	32	50
Smallest pulley diameter recommended, mm (ISO 4183)	20	50	75	125	200	355	500
Standard belt lengths, mm (ISO 4184)	200–500	400–1500	650–2700	950–6100	1550–10700	2750–15200	4650–16800

Figure 3-2. Examples of standardized classical V-belt sections with typical data (rounded figures)

3.3 Narrow V-belts

Narrow V-belts, also termed wedge belts, have a more narrow profile than the classical V-belts with a relative height of approximately 0,9, figure 3-3. The purpose of this construction is to obtain a more even distribution of

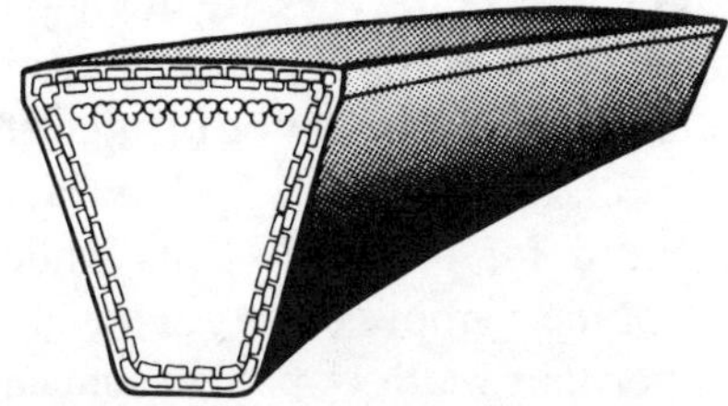

Figure 3-3. Principal design of narrow V-belt

tension forces between the reinforcing cords and thereby a higher power rating compared to classical V-belts of the same width. Tension forces on the reinforcing cords in a classical V-belt strive to pull the cords inwards to the bottom of the pulley groove unevenly with a consequent uneven stress distribution, figure 3-4. The central cords in the belt will thereby contribute less to the power transmission. The narrow V-belt section offers better support for all reinforcing cords and more even distribution of tension forces between them. The development of the narrow V-belt was accompanied by the introduction of improved reinforcement and rubber materials, cf figure 6-1, which also contributed to their improved power rating. As the weight (mass) of a narrow V-belt is much lower than that of the classical V-belt with the same power rating, higher belt speeds can be allowed, usually up to 35 m/s. The larger relative height of the narrow V-belt section increases its bending resistance compared to a classical V-belt of the same width and therefore requires larger minimum diameters of pulleys. In practice this is compensated

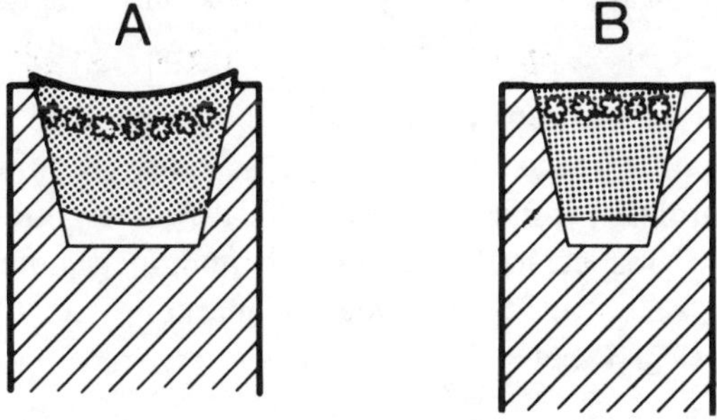

Figure 3-4. A classical V-belt (A) will deform more in the V-belt groove than a narrow V-belt (B)

for by the higher power rating of the narrow V-belts allowing smaller V-belt sections to be used that can tolerate smaller pulley diameters than classical V-belts with corresponding power rating.

Some important data for standardized narrow V-belt sections is presented in figure 3-5. The widths of sections SPZ, SPA, SPB and SPC are the same as for the classical V-belt sections Z, A, B and C. As both types are intended for use on pulleys with a groove angle of 32–38° the same pulleys may be used as for classical V-belt sections of corresponding width. This may be of interest in belt drives where a higher power transmission capability is desired. Earlier used classical V-belts may then be exchanged for narrow V-belts while retaining the old pulleys assuming the depth of their grooves is large enough. In multiple V-belt drives the power rating per unit width of pulley is much larger with narrow V-belts than with classical V-belts. This means a more compact design and lower moment on shafts and bearings.

The international standardization of narrow V-belts was complicated by the fact that it had to start from two well developed national standards, one made up by RMA in the USA and the other by DIN in Germany. The international standardization, cf figure 3-5, is mainly based on the DIN system. The RMA standard IP-22 is still of great importance especially in the USA. The three standard profiles 3V, 5V and 8V are intended for pulley grooves with a groove angle of 36–42°. Acceptable agreement and exchangeability is found between 3V and SPZ as well as between 5V and SPB. Section 8V is, however, larger than SPC. Profile SPC may in some cases be used in grooves intended for the 8V profile with some loss of power rating but the opposite is not recommended.

When V-belts are bent over the pulleys their thickness will increase at the middle of their sidewalls. To guarantee good, even surface contact in grooves, high efficiency and good power transmission capability the sections are often manufactured with somewhat concave sidewalls by use of specially shaped moulds.

Narrow V-belts were first developed to replace classical V-belts as fan belts in automobiles. For this application special, standardized sections are now available. ISO 2790 thus standardizes profiles AV10 and AV13, corresponding to sections SPZ and SPA. With respect to the advantages mentioned above the narrow V-belts have now taken an ever increasing market share of industrial V-belts from the classical V-belts.

Profile designation (ISO 4184)	SPZ	SPA	SPB	SPC
Profile dimensions, mm (DIN 7753)	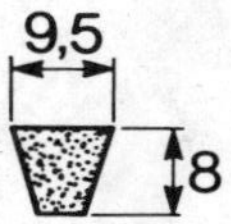 9,5 / 8	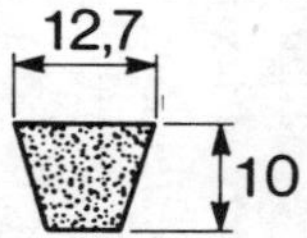 12,7 / 10	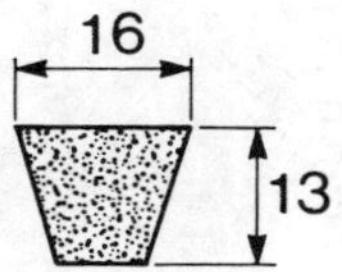 16 / 13	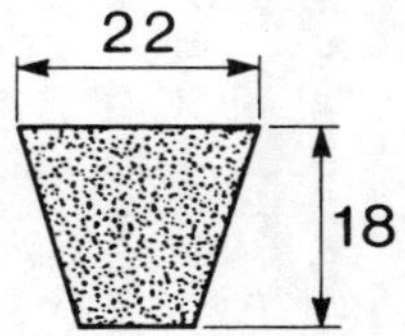 22 / 18
Max basic power rating per belt, kW (DIN 7753)	9,9	15	25	40
Smallest pulley diameter recommended, mm (ISO 4183)	63	90	140	224
Standard belt lengths, mm (ISO 4184)	650–3550	800–4500	1250–8000	2000–12500

Figure 3-5. Examples of some standardized narrow V-belt sections with typical data (rounded figures)

3.4 Cut V-belts

Cut V-belts, also termed raw edge V-belts, is a new V-belt type that has taken a considerable market share during the last years, figure 3-6. Cut V-belts have no textile wrapping and can thus be manufactured with a less labour-intensive process, which, however, limits belt length to approxomately 5 m, see chapter 7. The manufacturing methods also makes it simpler to produce belts with a cogged underside thereby reducing the bending resistance of the belts and allowing them to operate on pulleys with approximately 20% smaller diameters. When cogged belts are used on larger pulleys, however, their contact area in the grooves is somewhat reduced compared to non-cogged belts which may decrease their power transmission capability somewhat. At high belt speeds the cogging may also lead to a somewhat increased noise level. The decreased bending resistance reduces mechanical losses during bending and leads to improved efficiency and reduced working temperature. Heat dissipation will further be increased by the larger area between the belt and the surrounding atmosphere and by the air turbulence around the cogs during operation.

The change from wrapped to cut V-belts has, as with the change from classical to narrow V-belts, been accompanied by improvements in textile reinforcement and rubber materials, cf figure 6-1. The material development was necessary to compensate for the stabilizing and cohesive effect of the earlier used textile wrapping. Cut V-belts thus require a bottom rubber with improved resistance to compressive and shear forces to be transmitted from the sides of the pulley groove to the reinforcement as well as improved

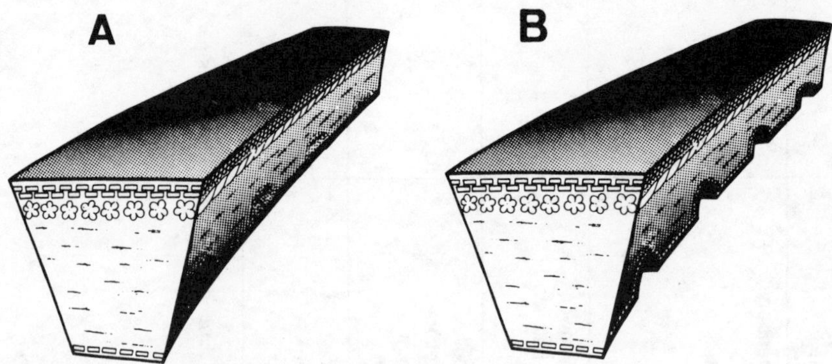

Figure 3-6. Principal design of cut narrow V-belt
A. With smooth, flat underside
B. With cogged underside

adhesion between various materials in the V-belt. During manufacture one of the cord strands is cut through leaving an open end at each sidewall on the belt. This additionally increases the requirements on good adhesion between reinforcement cords and rubber to avoid fraying at the edges. Elimination of the textile wrapping also means that more space is available for textile reinforcement, which allows increased power ratings within a given belt section.

The lighter and stronger construction of cut V-belts makes belt speeds of up to 45 m/s possible for cut V-belts with narrow V-belt sections and up to 35 m/s for cut V-belts with classical V-belt sections. This means considerably increased power ratings, compare for instance data in figure 3-7 with corresponding data for standardized classical V-belts in figure 3-2 and for standardized narrow V-belts in figure 3-5. As the cutting up of the individual belt sections takes place when the sleeve is vulcanized it will not cause any displacements of the cords. The absence of joints normally found in the textile wrapping will further contribute to smoother running.

Cut V-belts are normally manufactured in narrow V-belt sections but classical V-belt sections are also available. The first important use was as fan belts for automobiles but industrial applications are of growing importance.

During the manufacture of cut V-belts a certain concavity of the sidewalls is obtained. As was already mentioned above in the discussion about narrow V-belts this concavity is desirable as it compensates for the bulging out of the V-belts during bending and improves their contact with the groove sides.

3.5 Joined V-belts

Joined V-belts, sometimes also termed multi-V-belts, power bands or multi-band belts, can be described as 2–5 individual belts joined side by side by a reinforced rubber layer on the top side, figure 3–8. The term "multi" may cause some confusion as "multiple V-belt" implies the use of several individual V-belts running side by side in a belt drive. Joined V-belts with two sections were earlier often termed twin V-belts. The construction is developed mainly to cope with large pulsating and shock loads as with compressors, combustion engines, crushers, etc, where there is risk that individual belt sections may tilt and then rapidly get destroyed. Joined V-belts also give a

Profile designation	SPZX	SPAX	SPBX	ZX	AX	BX	CX
Profile dimensions, mm	9,7 / 8	12,7 / 10	16,3 / 13	10 / 6	13 / 8	17 / 11	22 / 14
Max basic power rating per belt, kW	13	23	45	3,1	9,2	16	28
Smallest pulley diameter recommended, mm	56	71	112	40	75	118	180
Standard belt lengths, mm	650–3350	750–3350	1250–3350	650–1900	650–3400	700–3400	1300–3350

Figure 3-7. Typical data (rounded figures) for cut V-belts with cogged underside of standardized classical or narrow V-belt profiles. Addition of "X" in the profile designation is used by many V-belt manufacturers to indicate cut V-belts (data for TrellPower X V-belts in catalogue from Trelleborg AB).

Figure 3-8. Principal design of joined V-belt

better stress distribution and more even slip between the individual V-belt sections. They are therefore well suited for the use of high-modulus and high-strength cord materials as, for instance, aramid cord.

Joined V-belts are manufactured with classical as well as narrow V-belt profiles, with or without textile wrapping. Joined V-belts without textile wrapping (cut belts) are also manufactured with cogged underside. Section sizes in the medium range are preferred. As the number of individual sections also can vary – usually 2, 3 or 4 – the number of variants of joined V-belts can be very large. International or national standards are not available and most joined V-belts are made on special order for a given application. It should be observed that the exchangeability between joined V-belts, with narrow belt section according to ISO 5290 and RMA IP-22, is limited. Examples of data for a number of typical joined V-belts are found in figure 3–9.

The reinforced rubber layer joining together the individual belt sections on their top side somewhat increases the height and bending resistance of joined V-belts. Too deep or worn pulley grooves should be avoided in pulleys for joined V-belts as there will be a risk of contact between the outer circumference of the pulleys and the joining layer on top of the belts. The profile sections in joined V-belts are therefore also manufactured with a somewhat larger height than the corresponding individual classical and narrow V-belt sections. The cord reinforcement in the joining layer is oriented transversely to the length of the belt in order to minimize bending resistance. Sometimes square-woven fabric is used for this application with the yarns placed at a 45° angle to the length direction of the belt.

3.6 Wide V-belts

Wide V-belts have a large width compared to their height (relative height around 0,3), figure 3-10. They are mostly termed variator belts, as their main use is in belt variators, i e belt drives with a continuously variable speed ratio,

Profile designation	3/5 V	2/3 V	2 HB	2 HC
Corresponding single profile	SPB/5 V	SPZ/3 V	B	C
Profile dimensions, mm	51 / 15	20 / 10	36 / 12	49 / 15
Number of single sections	3	2	2	2
Max basic power rating per belt, kW	60	20	13	26
Smallest pulley diameter recommended, mm	180	70	140	270
Standard belt lengths, mm	850–3150	530–1400	1500–4500	1500–5400

Figure 3-9. Examples of typical data (rounded figures) for joined V-belts (data for Multi-V-Belts from Trelleborg AB)

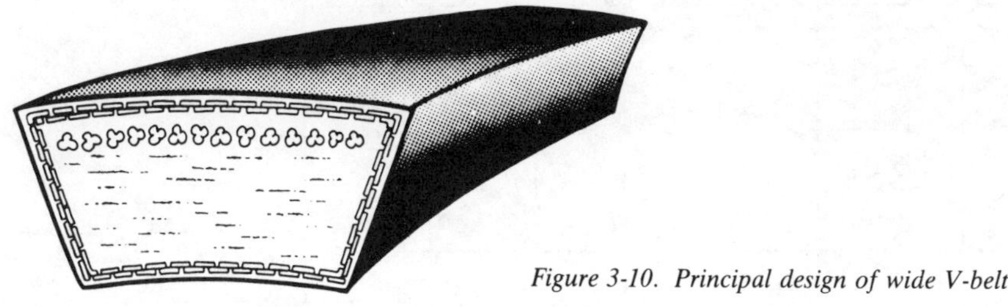

Figure 3-10. Principal design of wide V-belt

cf section 2.2. Classical V-belts may also be used as variator belts but the speed ratio range is then limited to 4,5:1. Increased speed range can either be obtained by increasing the width of the V-belts in relation to their height (lower relative height) or by decreasing the groove angle. Standardized variator belts according to ISO 1604 are manufactured for a groove angle of 26° while RMA IP-25 prescribes 22–36°. Depending on the construction of the variator asymmetrical V-belts may be preferred in special cases with for instance one side perpendicular to the top side and the other sloaping inwards. Wide V-belts can be manufactured with wrapped textile cover or as cut V-belts. The cut belt varieties are usually manufactured with a cogged underside so that the belts can run on small diameters. With belts of low relative height, cogged underside and small belt angles, speed ratio ranges of up to 9:1 can be obtained. Cogging also minimizes mechanical energy losses during bending, increases efficiency and reduces working temperature of the belts. Under these circumstances variator belts can operate at speeds of up to 45–60 m/s. This is utilized in automotive drive design while industrial or agricultural applications are usually limited to speeds of up to 40 m/s.

The transversal bending is larger for wide V-belts than for classical V-belts, cf figure 3-4. To counteract this tendency additional stiffening, reinforcing layers may be placed in the belt section. Another way to compensate for the bending action – applicable only to individually moulded, but not to cut, wide V-belts – is to manufacture them with a curved section, i e with a convex topside and a concave underside. The power transmission capability of variator belts is, with regard to the risk for bending and uneven stress distribution, smaller than would be expected from the large amount of reinforcing materials used.

Typical data for some standardized sections of wide V-belts are presented in figure 3-11.

Profile designation (ISO 1604)	W 25	W 31,5	W 40	W 50	W 63
Profile dimensions, mm (ISO 1604)	25,9 / 8	32,6 / 10	41,5 / 12,8	51,8 / 16	65,3 / 20
Max basic power rating per belt, kW (RMA IP-25 corresponding dimension)	2,4	4,4	8,4	14	17
Smallest pulley diameter recommended (cogged underside), mm (ISO 1604)	45	56	71	90	112
Standard belt lengths, mm (ISO 1604)	700–1600	900–2000	1150–2500	1450–3200	1850–4050

Figure 3-11. Examples of standardized wide V-belt profiles for industrial use with typical data (rounded figures)

3.7 Hexagonal belts

Hexagonal belts, also termed double-angle V-belts or duplex belts, can be described as two classical V-belt sections joined together on the their top sides, figure 3-12. They contain only one layer of reinforcement, placed in the middle of the section. The purpose of the construction is to get a belt that can drive on both sides in so-called serpentine drives, mainly used in agricultural machinery. Hexagonal belts are usually manufactured with textile wrapping. This, in combination with the large height of the belt section, increases its bending resistance and weight (mass). Consequently belt speed and power rating are limited as well as is the possibility of running over small diameter pulleys.

Figure 3-12. Principal design of hexagonal belt

Typical data for some standardized sections of hexagonal belts are given in figure 3-13. Belt sections are here designated by doubling the letters used for corresponding classical V-belt sections, e g AA for A, BB for B, etc. For agricultural belts an "H" is added to the belt designation, e g HAA, HBB, etc.

3.8 Circular section belts

Ropes and wires were the predecessors of V-belts. Wires and ropes may still be used for power transmission or braking, often utilizing the wedge effect by letting them run in a groove on a pulley or drum. Special circular section belts are available for power transmission, figure 3-14. They can be manufactured either homogeneous without reinforcement or with cord reinforcement in the centre. Their main use is for very low power ratings, e g in record

Profile designation (ISO 5289)	HAA	HBB	HCC	HDD
Profile dimensions, mm (ISO 5289)	13 / 10	17 / 13	22 / 17	32 / 25
Smallest pulley diameter recommended, mm (RMA IP-21)	80	145	235	345
Standard belt lengths, mm (ISO 5289)	1250–3550	2000–5000	2250–8000	4000–10000

Figure 3-13. Examples of standardized hexagonal belt sections for agricultural use with typical data (rounded figures)

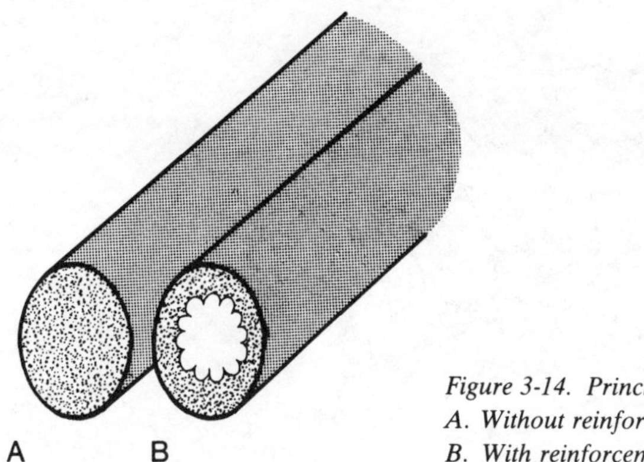

Figure 3-14. Principal design of circular section belt
A. Without reinforcement
B. With reinforcement

turntables. One advantage, which is sometimes of importance, is that the belt can, in drives with several pulleys, easily operate in several planes, i e "turn around corners".

3.9 Link belts

A special type of belt, also utilizing the wedge effect, is the link V-belt, figure 3-15. It consists of individual length sections of varying cross section joined together by mechanical link devices. The advantages of the construction are that belts of required length can easily be joined together in situ, that belt tension can easily be increased by taking out one length section of the belt,

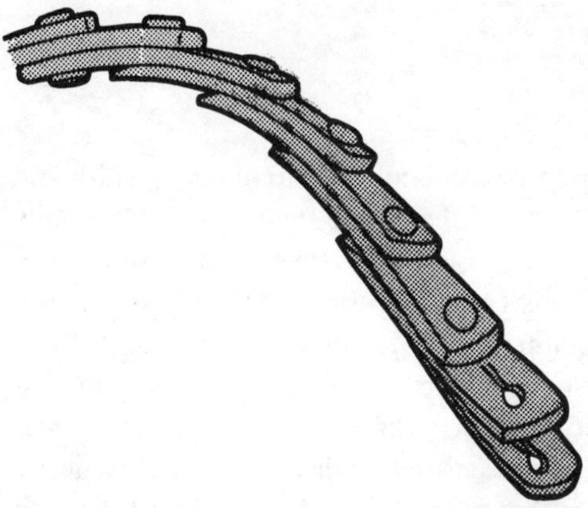

Figure 3-15. Example of link belt composed of fabric-reinforced rubber elements bolted together

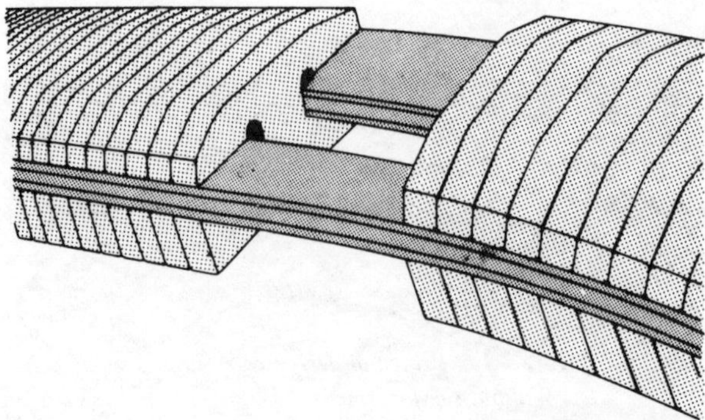

Figure 3-16. Variator belt made up of segments or lamellae for automotive drive. The belt contains flexible steel strips for load uptake on which trapezoidal-shaped steel lamellae are mounted. In operation the lamellae are pressed against each other and locked, whereby power can be transmitted by compressive force

and that repair can easily be made on the spot. As individual length sections can be manufactured with great accuracy, noise and vibration may be kept at a low level. Power rating is normally lower compared to belts with no mechanical links and the length stability of the belt poor (the belts will creep).

New types of link belts for variator drives have been developed, which are also termed segment belts or lamellar belts. They usually contain a thin, flexible load-carrying member for the power transmission, e g made from cord-reinforced rubber or a thin steel belt, onto which stiff, wedge-shaped lamellae or segments are threaded, mounted or cured. Various constructions have been suggested in which metallic components, textiles and rubber are combined in various ways, cf figure 3-16.

3.10 Flat belts

Flat transmission belts can be described as an endless strip of given width and thickness containing reinforcement of various types, figure 3-17. Originally they were manufactured from leather, but later textile came into use, impregnated and covered with various types of rubber, balata and plastics. Cotton fabric was the first textile material used for belt reinforcement. Belts were manufactured open-ended for mechanical joining, but woven endless constructions were later developed. Modern variants using cord reinforcement of various fibre types are manufactured endless. Modern varieties include reinforcement of oriented polyamide film (nylon) and with friction cover of chrome leather, rubber or plastics.

42

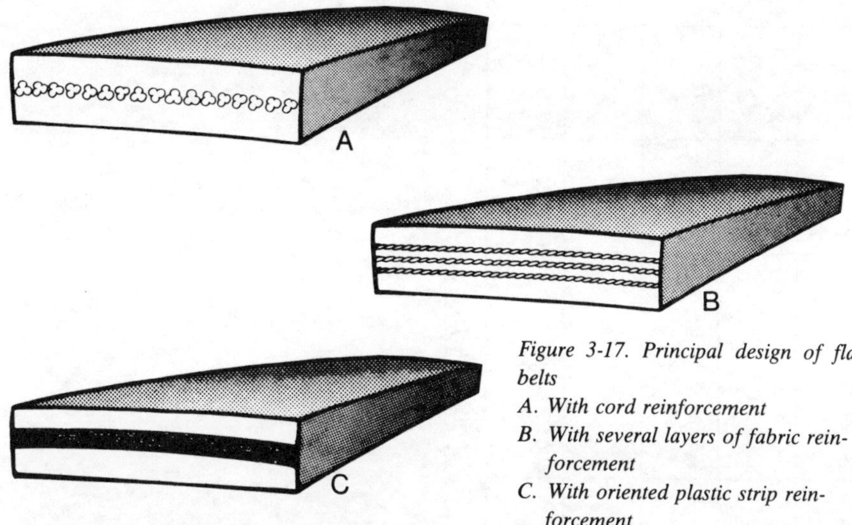

Figure 3-17. Principal design of flat belts
A. With cord reinforcement
B. With several layers of fabric reinforcement
C. With oriented plastic strip reinforcement

Flat transmission belts do not – contrary to all other types of transmission belts discussed above – make use of the wedge effect but rely solely on the friction between belt and pulleys for power transmission. To get adequate friction forces, belt tension must be kept high, which increases loads on shafts and bearings. A heavier and sturdier drive construction than for V-belt drives must therefore be used. This will deduct somewhat from the nominal high efficiency. The low height of the belt section means that small diameter pulleys may be used. Speed ratios of up to 20:1, even at low centre distances, are not unusual. The thin belt section further reduces mechanical energy losses during bending, lowering the belt working temperature, and limits centrifugal forces, making high belt speeds possible. The older, heavy types reinforced with cotton fabric and joined by mechanical fasteners allow speeds of up to 25 m/s, while modern endless types reinforced with oriented polyamide strip can be used at up to 100 m/s. As a rule flat transmission belts can run on both sides and are thus suitable for serpentine drives. The problem of keeping flat transmission belts on the pulleys can be solved by either using crowns or flanges, cf section 5.3.

Typical data for classical flat transmission belts of various thicknesses and normally with cotton fabric reinforcement is given in figure 3-18. Modern, lighter constructions utilizing improved materials allow much higher speeds and power ratings. It is thus sometimes stated that a surfacing of chrome leather can increase the coefficient of friction by a factor of 3. Flat transmission belts reinforced with oriented polyamide strip must be manufactured endless or joined in situ with the aid of special chemicals. The classical fabric

Type designation	190/40 190/60	240/40 240/60	290/40 290/60	340/40 340/60	385/60	425/60	450	500	560
Number of reinforcement plies (cotton fabric, ca 950 g/m^2)	3	4	5	6	7	8	9	10	12
Longitudinal tensile strength, kN/m width, min	190	240	290	340	385	425	450	500	560
Max basic power rating per 25 mm belt width, kW	4,6	6,1	6,3	9,2	10	12	14	15	18
Smallest pulley diameter recommended, mm	80	100	160	200	250	400	500	630	1000

Figure 3-18. Typical data (rounded figures) for some fabric-reinforced flat belts according to British standard BS 351

reinforcement has the advantage that it can easily be joined on the spot with the aid of mechanical fasteners. To increase power transmission capability either thicker belts are used with larger amounts of reinforcing materials or belt width is increased. Multiple belt drives, involving several individual belts running side by side in parallel as for V-belts, is usually not recommended for flat transmission belts. Examples can be found of belt drives where one single flat belt transmits up to 4000 kW.

3.11 V-ribbed belts

V-ribbed belts, also termed poly-V-belts, can be described as flat cord-reinforced transmission belts with lengthwise running triangular-shaped ribs with a top angle of 40°, figure 3-19. V-ribbed belts are thin and flexible and can drive on both sides. Ribs can be placed on only one or on both sides. V-ribbed belts are manufactured only as endless belts as they can't be joined in a simple manner. This construction leads to good support for all cords in the reinforcement and therefore an even load distribution is achieved. High-strength fibre materials may be used making light but still strong belts possible. This allows the use of very small-diameter pulleys. Mechanical efficiency is high and belt speeds of up to 50 m/s can be permitted.

V-ribbed belts have larger contact area between belt and pulleys than flat transmission belts of corresponding dimensions. According to classical friction law friction force, at a given normal load, is independent of the size of the contact surface, but practical experience is that V-ribbed belts can transmit higher frictional forces than flat belts of corresponding dimensions. This means that in a given power transmission lower belt tension can be used than with flat belts but this is still about 20% higher than for V-belts. This will reduce loads on shafts and bearings. V-ribbed belts track better than flat transmission belts but are not, however, quite as good as V-belts in this respect.

It should be observed that V-ribbed belts completely fill the grooves in the pulleys and do not wedge in the groove like V-belts. Increased frictional forces

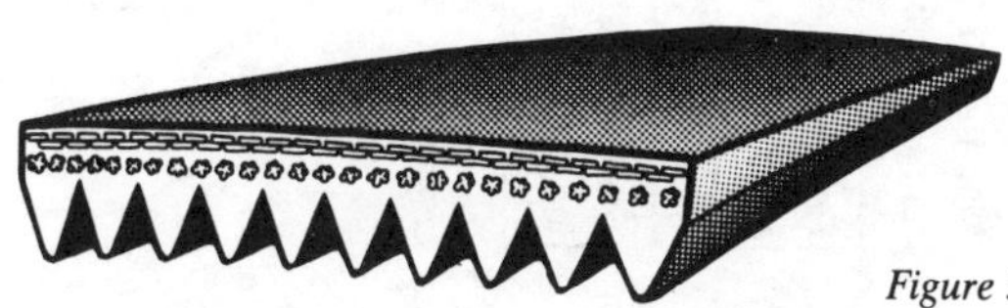

Figure 3-19. Principle design of V-ribbed belt

Cross section designation	PH	PJ	PK	PL	PM
Profile dimensions, mm					
Standard number of ribs	--	4–20	--	6–20	6–20
Max basic power rating per rib, kW	--	1,3	--	5,8	18
Smallest pulley diameter recom-mended, mm	13	20	40	75	180
Standard belt lengths, mm	--	450–2500	--	1250–3700	2300–9150

Figure 3-20. Typical data (rounded figures) for some standardized V-ribbed belt profiles according to American standard RMA IP-26

due to wedge action are not to be expected. Some types of V-ribbed belts have truncated or rounded ribs by which a certain wedging effect and an increased power rating is claimed. Belt drives with V-ribbed belts are sensitive to dirt which can fill up the grooves in the pulleys. Another drawback is that special, grooved pulleys have to be used.

Typical data for some standardized profiles of V-ribbed belts are given in figure 3-20. The smallest profile PH is used in miniature drives and the next larger PJ in household equipment drives. The profile PK is mainly used in automotive applications, while the larger profiles PL and PM find their use in industrial and agricultural applications.

3.12 Synchronous belts

Synchronous belts, also termed timing belts, cog(ged) belts or tooth(ed) belts, differ from all other types of belts in that power is not transmitted by friction but by the mechanical interlocking of cogs or teeth on the belt in corresponding recesses in pulleys, cf figure 2-7. This means that low belt tension can be used, which causes low loads on shafts and bearings. The principle is thus the same as for gear or chain drives. It gives accurate transfer of speed between driving and driven shafts. Synchronous belts are therefore mainly used where accurate transfer of speed in a drive is desired.

Synchronous belts can be described as flat belts with cogs or teeth on the underside, figure 3-21. As they have good flexibility they can also be used for drive on both sides. In this case the top side can either be cogged, see figure 3-22, flat or ribbed. Synchronous belts require reinforcement with low elastic and plastic deformation. High-strength and high-modulus fibres are therefore preferred, e g glass, steel or aramid fibres. As with flat or V-ribbed belts all cords are well supported over the width of the pulley, which facilitates the use of these reinforcing materials.

Classical types of synchronous belts have trapezoidal-shaped cogs, cf figures 3-21 and 3-22. A more even stress distribution can be obtained by using

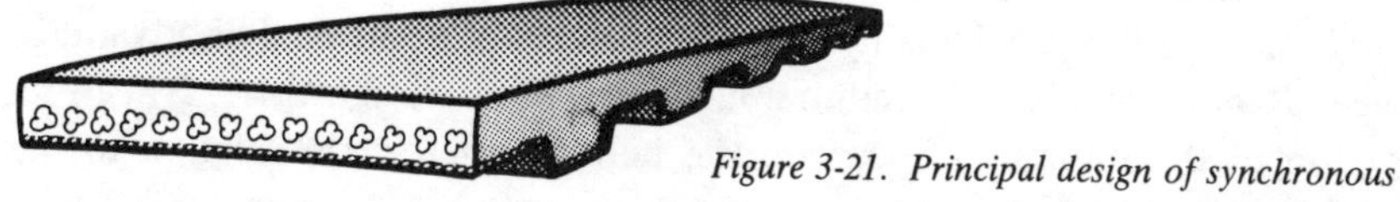

Figure 3-21. Principal design of synchronous belt

(rounded) cogs with the contour of a circular arc, figure 3-23, and these are therefore increasingly used. It makes higher power ratings possible and leads to a smoother and quieter drive. To increase the mechanical stability of the cogs and decrease their wear they are usually surfaced with polyamide (nylon) fabric, usually as two, crossed layers of polyamide cord fabric.

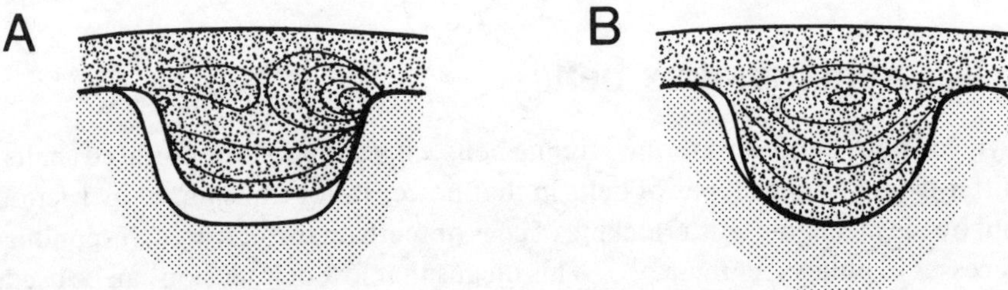

Figure 3-23. Stress distribution for various cog designs in synchronous belts
A. Trapezoidal-shaped cog
B. Circular section cog

Synchronous belts have a light and flexible section leading to high mechanical efficiency. They can be used on small pulleys with speeds of up to 60 m/s.

Typical data for some standardized profiles of synchronous belts are presented in figure 3-24. Belt lengths and pulley diameters must be chosen so that they correspond to integers of the pitch length. As for flat or V-ribbed belts, increased power ratings are obtained either by using belts of increased thickness, i e utilizing more reinforcement, or by increase in belt width.

Synchronous belts, with regard to their many advantages mentioned above, are increasingly used in many applications also where accurate speed transfer is not required. Their disadvantages are mainly to be found in the increased costs for belts and pulleys, the need for special pulleys and the higher demands on alignment during the installation of the drive. An additional disadvantage is the high noise level compared to non-cogged belts, cf figure 4-1. By careful selection of the number of cogs on the belt and pulleys in relation to the

Pitch code (ISO 5296, RMA IP-24)	XL	L	H	XH	XXH
Profile dimensions, mm (ISO 5296)					
Belt pitch, mm (ISO 5296)	5,1	9,5	12,7	22,2	31,8
Standard width, mm (ISO 5296)	6–10	13–25	20–75	50–100	50–125
Max basic power rating per 25 mm belt width, kW (RMA IP-24)	3,5	4,7	16	17	20
Smallest pulley diameter recommended, mm (ISO 5294)	16	36	63	125	220
Standard pitch lengths, mm (number of teeth) (ISO 5296)	150–660 (30–130)	315–1525 (33–160)	610–4320 (48–340)	1290–4445 (58–200)	1780–4570 (56–144)

Figure 3-24. Typical data (rounded figures) for some standardized synchronous belt profiles. Belts according to ISO 5296 and RMA IP-24 are fully interchangeable

rotational speeds of the shafts, careful adjustment of belt tension, accurate alignment of the drive and by surrounding the drive in a sound-deadening envelope the noise level can be reduced. The catalogues from the manufacturers give proper advice in this matter.

3.13 Special variants of V-belts

Besides the standard types of V-belts discussed above, which dominate the volume of the V-belt market, a great number of special variants exists, some of which are exemplified in figure 3-25. Among these the following can be mentioned:

Large belt angle: Some materials used in V-belts, e g special types of urethane rubber, have such a high coefficient of friction that V-belts with normal belt angle tend to get locked in the grooves by the wedging action and rapidly wear out or get torn off. Sufficient power ratings can then be obtained by using larger belt angles, usually around 60°, figure 3-25A. Here wear will be reduced and lower and lighter belt sections obtained, which permit belt speeds up to 50 m/s. Belts of this type were originally developed for automotive drives but due to their somewhat higher price they have not made a decisive breakthrough.

Lower pitch line: By moving the textile reinforcement somewhat deeper, under the top side of the belts, their flexibility and ability to run over pulleys of smaller diameters will increase, figure 3-25B. At the same time power ratings will be reduced as the available, smaller belt width will give less room for reinforcement. Belts of this type are of special interest in household equipment, which is characterized by low power ratings and compact design.

Additional reinforcement: Belts with added reinforcement are used in applications where large shock loads may appear, figure 3-25C. Textile layers on the underside of the belt will reduce the risk of cracking in applications, where the belts are double-bent, e g by the use of idlers on the outside of the belt, but will at the same time increase bending stiffness of the belt, figure 3-25D.

Belts made to slip: By selecting special materials with low friction, belts can be manufactured that slip under overloads, which may in special applications be to advantage.

50

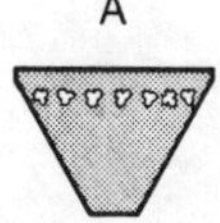 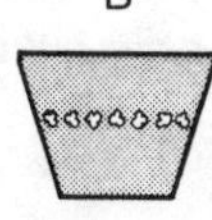  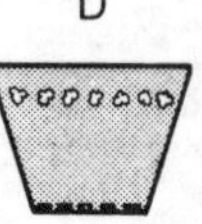 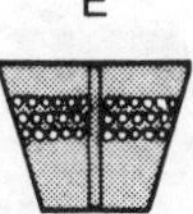

Figure 3-25. Some examples of V-belt of special construction
A. Wide belt angle
B. Low pitch line
C. Additional layers of reinforcement
D. Fabric-reinforced underside
E. V-belt for mechanical joining with fabric reinforcement and predrilled holes for mechanical fasteners

Belts for special environments: It is desirable in some applications that belts do not stain or discolour, e g in the food industry. The common rubber materials, black in colour, are then replaced by light-coloured ones. In oily environments belts manufactured from special oil-resistant materials must be used.

V-belts for mechanical joining: V-belts are normally manufactured endless to eliminate weak points in the construction and to use the reinforcement optimally. Sometimes the mechanical joining of V-belts on site is desired, for instance to allow installation on site or to shorten belt lengths in order to reach acceptable belt tension. The belts can then be supplied with pre-drilled holes for mechanical fasteners and reinforced with a fabric reinforcement that offers a better anchoring of the fasteners than cord reinforcement, figure 3-25E. Mechanically joined V-belts have lower power ratings and/or a shorter working life than endless V-belts.

V-belts with transport functions: V-belts are sometimes used for transport or forwarding of materials, thus excluding power transmission. Examples are transport of materials in agricultural equipment, continuous presses, etc. To facilitate feeding the top sides of the belts are usually given various profiled patterns, figure 3-26.

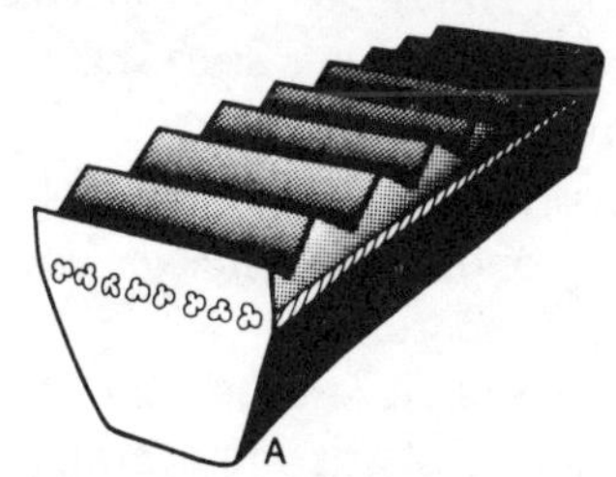 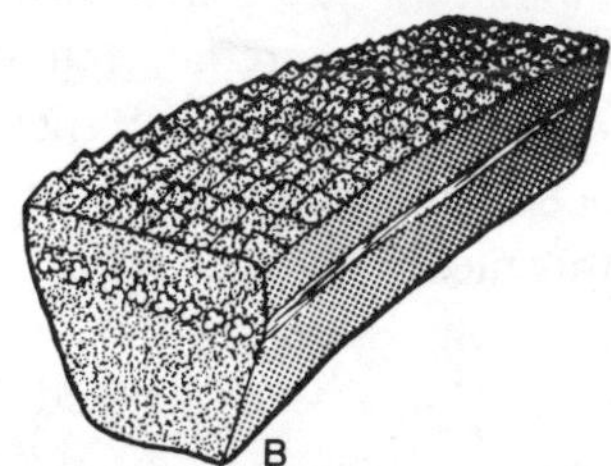

Figure 3-26. Examples of profile patterns on topside of V-belts with transport function
A. Transversal groove profile
B. Pyramid pattern profile

4. Requirements on transmission belts

4.1 Introduction

In this chapter the requirements on transmission belts will be enumerated first. Special requirements within the four main application areas will then be discussed. Next, the factors in belt design that bear upon these requirements are dealt with. Finally, comparative data for various belt types described in chapter 3 are tabulated. Where relevant, reference is made to following chapters where some of the problems are discussed in greater detail.

4.2 List of requirements

4.2.1 Power rating

The primary requirement on a transmission belt for use in a belt drive is that it can transmit a given desired power. This requirement is somewhat ambiguous as the power rating of a belt depends on many factors and is for instance directly proportional to belt speed. A given power can thus either be transmitted by combining a high torque with a low belt speed or a low torque with a high belt speed. At a given belt force difference between the strands of the belt the torque will increase with pulley diameter. The selection of a suitable combination of these factors will depend on the rotation speeds of driving and driven shafts and the need for a torque of a certain size. Where the option is free, high rotational shaft speeds are preferred in order to achieve optimum belt speed and maximum power transmission. For a given power to be transmitted this means cheaper electrical motors and smaller sheaves that can allow high rotational speeds without exceeding the permissible belt speed limits. Maximum permissible belt speeds mean lower belt tension which in turn leads to smaller belt dimensions and lower loads on shafts and bearings. The interaction between these various factors will be discussed in more detail in chapter 16.

4.2.2 Shock loads

Most belt drives are, in addition to normal loads, exposed to occasional overloads. These will for instance occur during starting up of the drive or with equipment with pulsating loads such as crushers, compressors or combustion engines. In the design calculation of belt drives this is taken into account by introduction of so-called service factors, cf sections 16.10 and 18.4. Shock loads are best absorbed by belts with high elastic extensibility, which, however, will negatively influence their dimensional stability. Shock loads can also be absorbed by allowing occasional belt slip. In exceptional cases even belt rupture will be accepted, as the belt usually is the cheapest and most easily replaceable part in the belt drive.

4.2.3 Working life

The more a belt is loaded, the shorter its working life will be, cf chapter 15. Therefore when designing a belt drive for a given power transmission, the desired working life of the belt should be indicated. At the expense of a shorter working life installation costs can be reduced somewhat by under-dimensioning. As is discussed in more detail in sections 4.3 and 15.8 requirements on working life may vary from application to application.

4.2.4 Speed ratio

Many belt drives require a certain speed ratio between the rotation speeds of the driving and driven shaft. Most transmission belt types offer acceptable options within wide limits in this respect. The use of pulleys of different diameter and the use of electrical motors of different rotation speeds will additionally increase the options.

4.2.5 Shaft centre line distance

From space and design considerations it is often desirable that the shaft centre line distance can be chosen within a wide range. The availability of a great number of belts of standard lengths or lengths made to order usually gives a wide choice in this respect.

4.2.6 Accuracy in speed transfer

In many belt drives, usually characterized as synchronous drives, great accuracy in speed transfer is desired, i e an almost exact rotation speed of the driven shaft in relation to the rotation speed of the driving shaft. This can be obtained with cogged or teethed belts working with mechanical interlocking, so-called synchronous belts, while belts transmitting power by frictional forces, such as V-belts or flat transmission belts, have a small and varying degree of slip making synchronization impossible, cf section 17.2.

4.2.7 Space requirements

Compact design of a belt drive is often highly desirable, e g in automobiles. Fulfilment of this requirement involves short shaft centre line distances, small diameter pulleys and belts of small section but still with acceptable power ratings.

4.2.8 Driving on both sides of belt

In belt drives including several driven shafts of reversed rotation it is imperative to use belts that permit driving on both sides without an unacceptable shortening of the working life. Belt drives of this type are often termed serpentine drives, cf figure 2-4.

4.2.9 Loads on shafts and bearings

In belt drives which transmit power by frictional forces, the belts have to be pressed against pulleys to generate the necessary levels of frictional force. At the same time it is desirable to keep the belt tension as low as possible in order to save shafts and bearings. Belt forces can be kept low by using rubber materials with a high coefficient of friction and by taking advantage of the wedge effect, cf chapter 12. With synchronous belts, where power is transmitted by mechanical interlocking, belt tension and thereby the load on shafts and bearings can easily be kept at a minimum.

4.2.10 Noise and vibrations

With the ever increasing demands on working environments the requirements on low noise and vibration levels become increasingly important. Compared to most other types of mechanical power transmission, belt drives offer excellence in this respect as the soft and elastic materials in the belts tend to dampen vibrations and prevent the generation and propagation of sound. The perceptible noise is air-borne while structure-borne sound cannot propagate via the belts. Noise levels usually increase with speed, figure 4-1, and with

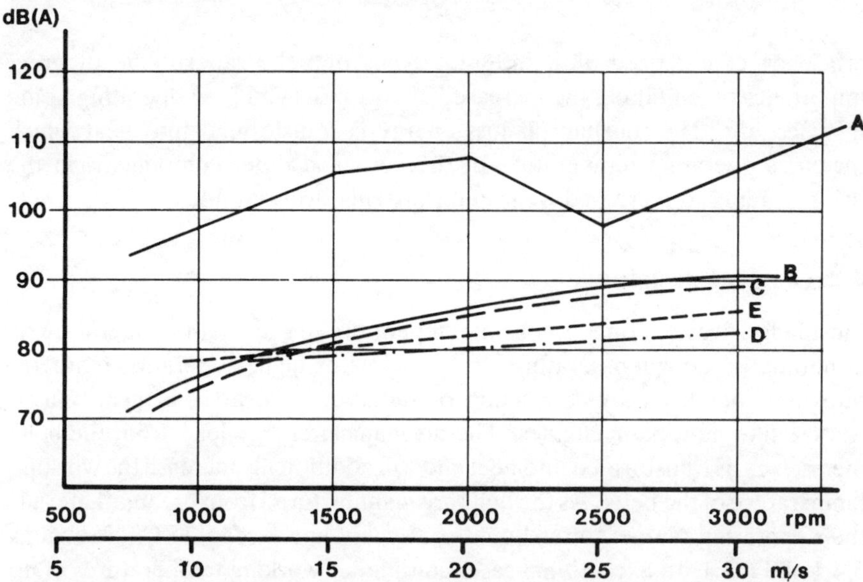

Figure 4-1. Examples of experimental data on noise level for belt drives measured at 0,5 m distance as a function of pulley rotation speed (rpm) or belt speed v (m/s) at power ratings in the range 20–70 kW and with the following belt types (Berg, 1977)
A. Synchronous belt   *D. Classical V-belt*
B. Flat belt   *E. Narrow V-belt*
C. Flat belt with idler pulley

the loading of the drive. Belts with varying section, for instance derived from manufacturing tolerances or with cogged underside, tend to give higher noise levels than belts of even section. Belts that undergo heavy slip, let us say more than 10–20%, for instance when starting up the drive or exposing it to heavy overloads, can give rise to a squeaking noise. Different investigations present conflicting data on the noise levels to be expected by the use of various types of belts. Practical testing is recommended to find the best solution in an important application. Vibrations in V-belts are discussed in more detail in section 19.9.

4.2.11 Mechanical efficiency

During bending, stretching and slip of the belts at the pulleys mechanical energy is consumed which is not completely recovered during elastic deformation, but partially dampened and transformed into heat. Belts will therefore operate with a certain mechanical efficiency, usually within the interval 90–97%, cf chapter 17. The relatively small differences in mechanical efficiency experienced with different types of belts can still be of great importance when taking the increased energy costs and long operating hours into account. The mechanical loss energy is transformed into heat which increases operating temperature of the belts and other components in the drive, leading to increased wear and shortened working life.

4.2.12 Heat resistance

Most belt drives operate at normal temperature, e g at room temperature or at normal outdoor temperature. In special cases higher operating temperatures are found, e g by convection or radiation of heat from nearby heat sources like motors or engines. The mechanical energy lost, from the belts themselves, is transformed into heat and will additionally increase the working temperature of the belts. As the belts are manufactured from organic materials their operating temperature should preferably not exceed 70°C, cf sections 6.4.1 and 6.5.1. In exceptional cases continuous working temperatures of up to 100°C may be permitted. By selecting special, heat-resistant reinforcement and rubber materials working temperatures of up to 150°C may be used with the sacrifice of some other properties, but above this temperature organic materials are generally not recommended. High temperatures can be avoided by good ventilation, shielding-off or elimination of radiant heat sources and by the use of belts with high mechanical efficiency and thereby low heat generation. An increase in the surrounding temperature of 20°C, roughly corresponds to an increase of 10°C in the operating temperature of the belt, which, according to section 6.4.1, cuts the working life roughly in half. To the requirements on heat resistance can be added requirements on low flammability to prevent fire from spreading along the belt, from the point of ignition, and requirements on low electrical resistivity, to prevent ignition from spark generation caused by electrostatic charge built-up.

4.2.13 Resistance to chemicals and environment

Chemical factors in the environment can have a deleterious influence on the belts and require that special belt materials be used. In oily atmospheres, e g involving oil splash or oil mist, special oil-resistant materials have to be used. Various solvents may in the same way require the use of rubber materials with good resistance to the solvent in question. The resistance of rubber in general to acid or alkaline solutions is usually better than that of other materials in the belt drive and therefore requires no special attention. High atmospheric ozone concentrations, as occur in the neighbourhood of electric discharges or heavy electric machinery, may require rubber materials of especially good ozone resistance. Most rubber materials, used on the outside of V-belts, are selected to offer acceptable ozone resistance to normal, atmospheric ozone concentrations, cf section 6.4.1.

4.2.14 Installation

Simple installation of V-belts, including endless belts, can be achieved by applying normal routines, cf section 9.2. In areas of difficult access and where suitable standardized belt lengths are not available mechanical joining on site may exceptionally be used, cf section 9.4. Mechanical joints will form a weak point in the construction, however, making efficient utilization of the reinforcement difficult and shortening the working life of the belts. Good parallelism of shafts and placement of pulleys in the same plane is usually aimed at during installation. The requirements on exactness vary between different types of belts.

4.2.15 Maintenance

Belt drives require no lubrication, as do drives containing metallic power transmission elements, which simplifies maintenance. On the other hand belts will suffer some increase in length during operation, which may require correction of belt tension to avoid slipping, cf section 9.3. Various devices are also available to keep belt tension constant, irrespective of any permanent deformation, cf chapter 14.

4.2.16 Spare part availability

Belts and pulleys are well standardized components, cf chapter 10, and are usually stock items at belt manufacturers and machine equipment suppliers. Spare part service is therefore excellent but may be more problematic when made-to-order, special belts are used.

4.2.17 Economy

After all the technical requirements have been met, the economy of the belt drive will become the decisive factor for the final choice. The costs shall not only include the purchase price of belts and pulleys but also the installation and maintenance costs and expected useful working life if this is not included from the beginning. The costs of energy losses, taking the mechanical efficiency of the power transmission into account, should also be considered.

4.3 Requirements in some important application areas

4.3.1 Industry

Industry represents the first application area for power transmission belts and is still the most important one, cf a modern example in figure 4-2. The market share is roughly 40%, of which half is for original equipment and half for replacement. Requirements vary from application to application but the most typical ones are:

- High power transmission capability, usually requiring multiple belt drives with belts of fairly heavy sections
- Long working hours, often continuous operation, requiring a long working life, often in the order of 25 000 h
- Most typical are simple drive designs involving one driving and one driven shaft
- High reliability and even quality

Figure 4-2. A rotary cutter, with a power rating of 200 kW, is driven by an electric motor with the aid of a multiple belt drive, including 14 narrow V-belts of SPC section, at a speed ratio of 2,5:1 and a belt speed of 19 m/s

Figure 4-3. Agricultural belt drives, in this case in a harvester, are often characterized by a great number of driven shafts which may require several belts or one belt driving several shafts

4.3.2 Agriculture

Applications involve various types of agricultural machinery, the harvester being a typical example, figure 4-3. The market share is roughly 15%. The following conditions are characteristic:

- The use of variator drives is common
- Several driven shafts are often included, some with reversed directions of rotation, the so-called serpentine drives
- The equipment is typically utilized during a limited period, e g during the harvesting season, meaning fairly low requirements on working life, usually in the order of 500–3 000 h
- Dusty and dirty operating conditions

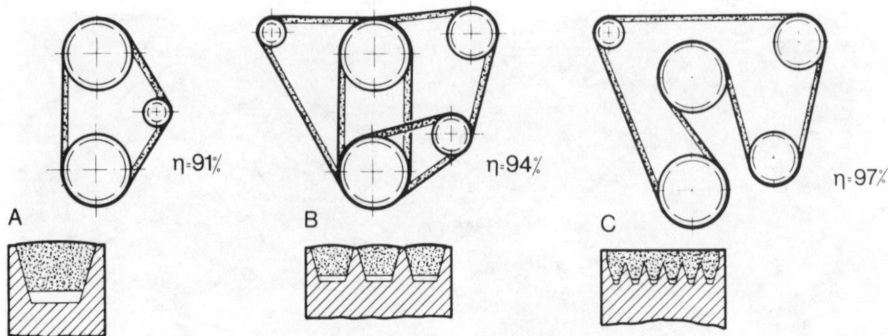

Figure 4-4. Steps in the development of automotive belt drives with corresponding improvements in efficiency η
A. Single, classical V-belt
B. Several narrow V-belts driving different units
C. One single V-ribbed belt used for all units while driving on the inside as well as on the outside (serpentine drive)

4.3.3 Automobiles

V-belts were originally developed for use in automobiles as so-called fan belts. The development of new improved V-belt types have since been initiated by increasing demands from the automotive industry, figure 4-4. The market share is about 35%. The following requirements are now typical, cf figure 4-5:

– The driving shaft must transmit power to several driven shafts for various pieces of equipment such as fan, generator, water pump, air-conditioning equipment, servo-hydraulic units, etc, which often requires several belts or serpentine drives. Many modern automobiles have direct drive for the thermostat-actuated fan, which makes the term "fan belt" incorrect.

– The increasing number of auxiliary equipment to be driven have increased the power requirements from originally a few kW to more typically 10–25 kW today. As an example, the Volvo PV 444 from the early 1950's with a 38 kW engine had one single V-belt to drive the generator, water pump and fan with design power of 2,6 kW, while a Volvo 760 from the early 1980's with a 115 kW engine utilizes four V-belts, of which two, running in parallel, drive the generator, water pump and fan with a design power of 8 kW, one drives the air-conditioning unit at a design power of 8 kW and one drives the servo-hydraulic unit at a design power of 3,5 kW.

– The increasing demands on reliability and freedom of maintenance in automobiles, i e long time between service, require that belt drives should not need service, e g repeated adjustment of belt tension.

Figure 4-5. In a 12-cylinder engine, on a luxuary car, all units including generator, fan, water pump, servo-steering aggregate, servo-braking aggregate, air conditioning, etc, are driven by three V-belts from the crankshaft. The most difficult problem in manufacturing these belts, has been to control the hot-stretch conditions for the reinforcement used, so that belts will neither creep nor shrink during the large temperature variations experienced with this engine.

– Operating conditions in the engine compartment involve radiant heat, oil splash, dust and dirt.
– High requirements on compact construction, low price and weight assume a compromise with working life requirements which usually lie within the interval 1 000–10 000 h.

4.3.4 Home appliances

Many home appliances require belt drives, for instance dish washers, washing machines, mixers, sewing machines, kneaders, electric hand tools, fans, pumps, oil burners, lawn mowers, etc, figure 4-6. The market share is about 10%. Typical requirements are:

– Low power ratings meaning that small belt sections and single belt drives can be used

Figure 4-6. Single V-belt drives with belts of fairly small section are normally used in household washing machines

- Intermittent operation with many starts and stops, the operating life requirement usually being 500–2 000 h
- Low noise level and compact design

4.4 Belt performance versus belt construction

4.4.1 Materials

The development of improved belt types has always been accompanied by the use of improved materials, cf figure 6-1. The power ratings of belts have thus increased while keeping the original belt section or alternatively allowing the use of smaller belt sections to transmit a given power.

4.4.2 Adhesion between components in the belt

In order to take advantage of improved textile or rubber materials it is absolutely necessary that they can be firmly bonded together. Poor adhesion between the components in a transmission belt will rapidly lead to separation failures and finally to complete failure of the belt. The technology needed to achieve good adhesion between the components in a belt is as important as the technology for developing improved materials, cf section 7.3.

4.4.3 Even stress distribution

Constructions based on organic materials, such as textiles, rubbers or plastics, are very sensitive to stress concentrations, which can lead to rapid failure or initiate cracks that gradually will lead to fatigue failure, cf chapter 15. Maximum life and optimum utilization of materials require that stresses be distributed as evenly as possible. A problematic case is illustrated in figure 3-4, demonstrating how some of the cords in V-belts of classical or wide section are forced down radially in the pulley groove. This will lead to an uneven stress distribution in the reinforcement and consequently lower power rating or, alternatively, a shorter working life. The problem can in this case be overcome by use of narrow V-belt sections or by use of flat or V-ribbed belts where the cords are better supported and a more even stress distribution is obtained. Another example of uneven stress distribution is to be found on the cogs of synchronous belts, cf figure 3-23. A more even stress distribution can here be obtained by using cogs of another shape than the traditional trapezoidal one. Many additional examples could be cited to demonstrate how belt constructions have been perfected by learning from practical experience. Small modifications in construction or dimensions can lead to large improvements in power ratings or belt working life.

4.4.4 Thin and light section profile

During the development of new, improved belts the application of improved material, improved adhesion and improved constructions, has made possible thinner and lighter belts for the transmission of a given power. This leads to the following advantages:

- The increased flexibility of the belts makes the use of smaller pulleys possible leading to reduced space requirements and the possibility of higher shaft rotation speeds.
- The mechanical loss of energy during bending of the belts over pulleys will be reduced, leading to increased mechanical efficiency of the belt drive and

less heating of belts when running. The smaller section of the belts will further facilitate the dissipation of heat, reducing belt working temperature and increasing working life.

– The smaller and lighter belt section means lower mass and lower centrifugal forces making higher belt speeds possible and consequently higher power ratings or, alternatively, lower loads on shafts and bearings at the same power rating.
– Lower cost.

4.5 Comparison of performance of various belt types

Figure 4-7 summarizes and compares the performance of various belt types. The wide range of variants and sizes for each belt type and the difficulty in quantifying properties in an objective manner tends to make comparisons of this type arbritary. It is hoped, however, that the table may serve as a rough summary of the contents of this and preceding chapters.

Belt type	Power transmitted by interlocking (I) or friction (F)	Wedge effect	Load on shafts and bearings	Belt angle,°	Relative belt height	Max (optimal) belt speed v m/s	Max speed ratio c_p	Max flexural frequency Hz	Efficiency η at full load %	Can drive on both sides of belt	Max power rating P at 25 mm belt width kW	Grading: 1=excellent, 2=good, 3=moderate, 4=poor				
												Quietness	Accuracy in speed ratio	Compact design	Simple installation	Acquisition economy
Classical V-belts — with textile wrapping, uncogged underside	F	yes	moderate	36–38	0,7	25	7:1	30	89–94	no	15	1	2	3	2	2
— without textile wrapping, cogged underside	F	yes	moderate	36–38	0,7	35	8:1	50	91–96	no	30	2	2	2	2	2
Narrow V-belts — with textile wrapping, uncogged underside	F	yes	moderate	36–38	0,9	35	7:1	50	91–96	no	40	1	2	2	2	1
— without textile wrapping, cogged underside	F	yes	moderate	36–38	0,9	45	8:1	90	93–98	no	70	2	2	1	2	1
Joined V-belts	F	yes	moderate	36–38	–	20–40	7:1	40	89–94	no	70	1	2	2	2	2
Wide V-belts (variator belts)	F	yes	moderate	22–36	0,3	40–60	9:1	50	89–94	no	20	1	2	2	2	2
Hexagonal belts	F	yes	moderate	36–38	1,4	20	7:1	30	87–92	yes	15	1	2	3	2	2
Wide angle V-belts	F	yes	moderate	60	0,7	50	10:1	100	91–96	no	25	1	2	2	2	2
Link V-belts	F	yes	moderate	36–38	0,7	20	7:1	50	91–96	no	10	1	2	2	1	2
Circular section belts	F	yes	moderate	–	–	25	15:1	40	90–95	yes	2	1	3	3	1	2
Flat belts — fabric reinforced, older type	F	no	high	(180)	0,05–0,2	25	15:1	50	91–96	yes	20	2	2	3	2	2
— cord or oriented plastic strip reinforced, modern type	F	no	high	(180)	0,03–0,1	50–100	20:1	150	93–98	yes	70	2	2	2	2	2
V-ribbed belts	F	no	moderate	(180)	0,1–0,3	50	15:1	100	92–97	possible	50	2	2	1	3	3
Synchronous belts	I	no	low	(180)	0,1–0,3	60–80	15:1	200	93–98	possible	40	3	1	1	3	4

Figure 4-7. Comparative data for different belt types. The magnitude of the figures will depend strongly on conditions selected and requirements applied. The figures are only given here to indicate typical differences between various belt types

5. Pulleys

5.1 Introduction

The main components of a belt drive are belts and pulleys, also, especially in American literature, called sheaves. In this chapter pulleys will be discussed, primarily those for use with V-belts. Pulleys for flat belts, V-ribbed belts, synchronous belts and variators will also be briefly discussed. Various ways of characterizing pulley groove dimensions for V-belt drives, are discussed in section 11.2.

5.2 Grooved pulleys for V-belts

5.2.1 Groove design

A pulley for use with V-belts can be described as a wheel or drum with circumferential grooves, the number and dimensions of which shall fit the V-belts used in the drive, figure 5-1. Dimensions of pulleys must, with regard to this requirement, be well standardized. The hub in the centre of the pulley shall allow firm attachment on driving and driven shafts.

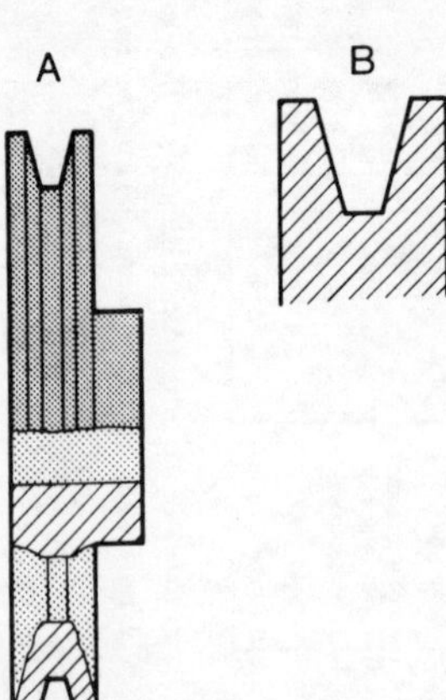

Figure 5-1. Sketch of V-belt pulley
A. Section of the whole pulley
B. Section of the pulley groove

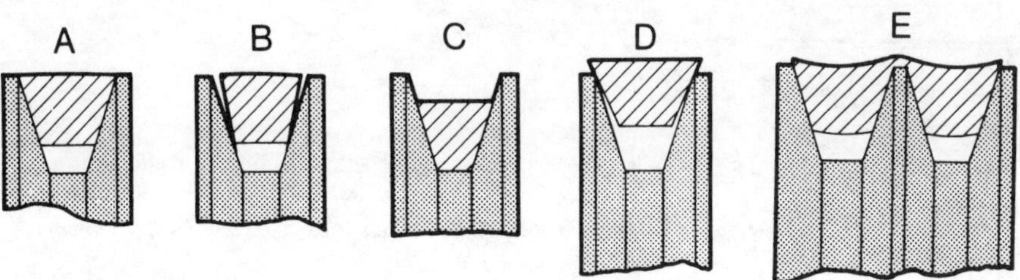

Figure 5-2. Examples of misfit between belts and pulley grooves
A. Correct fit
B. Deviation between belt angle and pulley groove angle
C. The belt bottoms in the pulley groove, whereby the wedge effect is lost
D. The belt rides high in the groove leading to lowered power transmission and risk of tilting
E. Joined V-belt may not rest on top of ridges of pulleys

Dimensions of the pulley groove are adapted to the dimensions of the V-belts used. The angle formed between the prolonged side walls of the groove is termed angle of the pulley groove. For good contact between V-belt and groove side-walls the angle of the pulley groove shall agree with belt angle. The depth of the pulley groove should furthermore be suited to the height of the belt section. The width of the grooves is usually chosen so that the top surface of the belt will be in line with the outer circumference of the pulleys. If a belt protrudes too far above the pulley circumference (ride-out) there is a risk that it may run off the groove, tilt and rapidly get destroyed, especially in drives with large shock loads. In such applications special pulleys with deeper grooves are sometimes preferred. It is evident that grooves should not be so shallow that the belts rest on their bottom, whereby the wedge effect is lost and power transmission capability drastically reduced. Deep pulley grooves should not be used with joined V-belts as, even with correct groove width, there is a risk that the adjoining layer in this belt construction will come into contact with the circumference of the pulleys. Wear of belts and pulley grooves can also lead to the pre-requisites for a good belt drive, as discussed above, being spoiled. Some examples of the effects of inproper design are illustrated in figure 5-2.

During the bending of a V-belt over a pulley its section will be deformed so that the belt angle will somewhat decrease (up to 6–12° for small pulleys), figure 5-3. To compensate for this and maintain good contact, pulley groove angles are usually somewhat smaller than the belt angles. The difference will become larger, the greater the bending is, i e larger with small diameter pulleys. The difference is usually 2–6°, but the exact figures can be found in relevant standards.

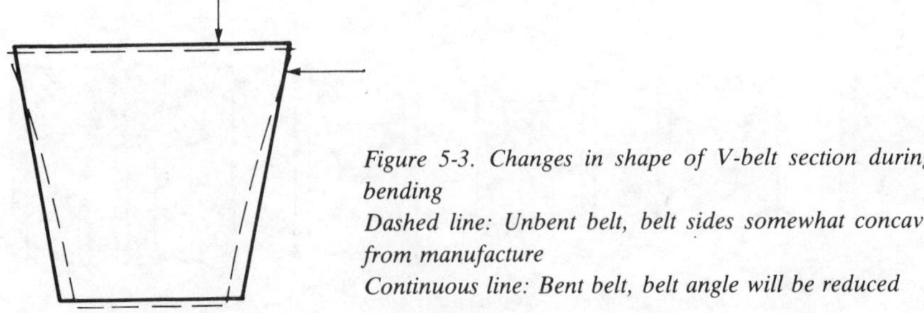

Figure 5-3. Changes in shape of V-belt section during bending
Dashed line: Unbent belt, belt sides somewhat concave from manufacture
Continuous line: Bent belt, belt angle will be reduced

To check the groove angle a simple maximum and minimum template can be used, figure 5-4, from which incorrect angles can be read from the play of the template.

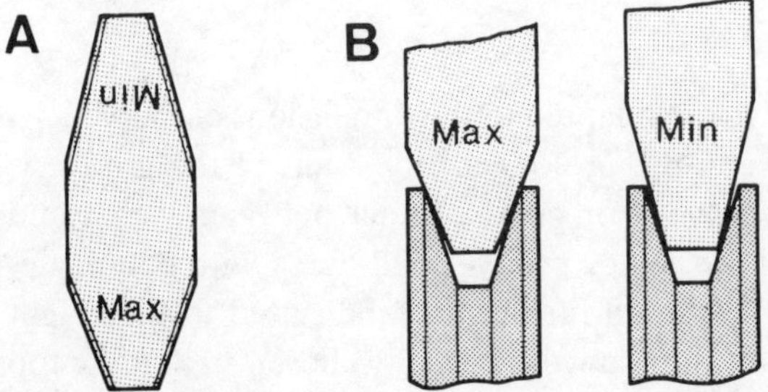

Figure 5-4. Limit gauge for determination of groove angle according to ISO 255. The gauge contains two templates with maximum and minimum permissible deviations from the nominal groove angle, in this case 38°
A. ISO limit gauge
B. Placement of limit gauge in the groove to be checked

5.2.2 Diameters

The width of the pulley groove and the diameter of the pulley can be defined in different ways, cf section 11.2. Standards for pulleys give a choice of various diameters for given pulley groove dimensions, cf Appendix E. The selection of a suitable diameter should be made with regard to desired torque, rotation speeds of shafts, maximum permitted belt speed and minimum recommended pulley diameter for a given belt section. The operating diameter (pitch diameter or datum diameter, cf section 11.2) can be checked with a simple depth gauge, figure 5-5 A. This gauge registers the distance between the outer diameter of the pulley and the depth of the template in the groove. Accuracy in measurement is based on the assumption that the outer diameter of the pulley, measured separately by a suitable instrument, is perfectly cylindrical. Alternatively another method of measurement is used, figure 5-5 B. Two steel balls or cylindrical rollers of accurate dimensions are placed diametrically

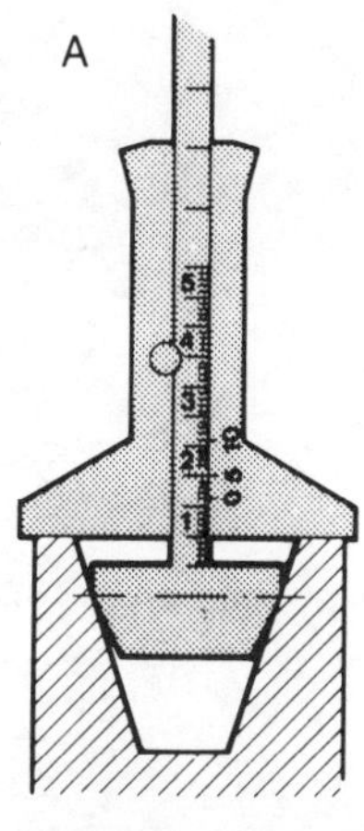

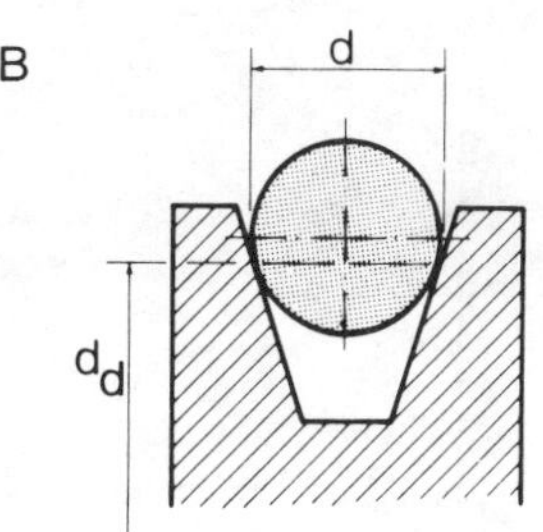

Figure 5-5. Inspection of datum diameter of pulley according to ISO 255
A. With depth gauge starting from the external diameter of the pulley
B. With steel balls or cylindrical rollers placed in the pulley groove

opposed in the pulley groove and the distance between the tangential planes to their protrusions is measured. The diameter of the balls or cylindrical rollers is chosen according to the dimensions of the groove to be checked. Compare relevant standards which also indicate how pulley diameter is calculated from measured data.

In cases where step-wise variation of speed ratios is desired, so-called stepped pulleys are used, figure 5-6, often manufactured from aluminium.

Figure 5-6. Stepped pulley for stepwise variation of speed ratio and rotation speed by moving the belt between the grooves in the pulley. The counterpulley on the other shaft is similar but with pulley diameters in reversed order so that belt length can be kept constant when moving between grooves

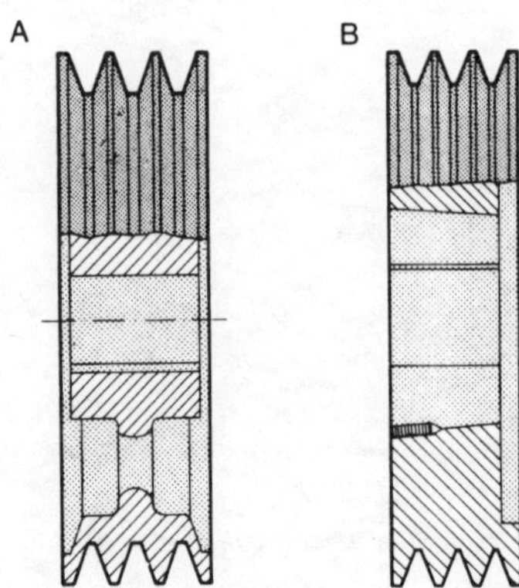

Figure 5-7. Common hub design principles applied to V-belt
A. Integral hub
B. Removable hub

5.2.3 Hub and shaft connections

The hub in the centre of the pulley shall allow firm mounting of the pulley on the shaft. It may either be an integral part of the pulley (straight-bore pulley), figure 5-7 A, or a removable hub bushing (bushing pulley), figure 5-7 B. A hole is drilled in the centre of the hub to fit the diameter of the shaft. Certain pulleys with pre-drilled holes may be on stock, but most often holes are drilled to order to fit the wide variety of shaft diameters. Pulleys with integral hubs are mostly fastened on shafts with the aid of keys and key-ways while pulleys with removable hub bushings may either be fastened by keys, screws or by clamping on the shaft. Removable hub bushings allow quicker mounting and dismounting than do pulleys with integral hubs.

5.2.4 Materials and construction

Common types of pulleys are:

Cast iron pulleys with integral hubs: This is the most common type of pulley for industrial belt drives, figure 5-8, and is manufactured from cast iron. They are often termed "standard pulleys". Their integral hub is mounted with keys and key-ways on the shaft. They are available in all standard diameters for standard V-belt sections and with one or several pulley grooves. Small diameter pulleys usually have an unbroken web while larger diameter ones have spokes. Cast iron has good corrosion resistance and is to some extent sound-deadening. Reliability, precision and service life are also good. For pulleys operating at very high speeds steel is often preferred to cast iron.

70

*Figure 5-8. Cast iron pulley
with integral hub with keyway
for mounting on shaft*

Cast iron pulleys with removable hub bushings: These pulleys are manufactured from the same materials and machined in the same way as standard pulleys. Their prime advantage – the simple mounting and dismounting – must be paid for by a higher price. An example of design is given in figure 5-9.

Aluminium pulleys: These pulleys are manufactured from aluminium either by casting or machining, figure 5-10. They can be delivered either with integral hubs or dismountable bushings. For high power ratings, where the soft aluminium cannot take the loads involved, the bushings can be manufactured from steel. Normally aluminium pulleys are intended for low power ratings. Their advantages are primarily their light weight, good corrosion resistance and (at least for small diameter pulleys) low price.

Stamped steel plate pulleys: Pulleys made from stamped steel plate, can only be designed with one or two grooves, figure 5-11. They are primarily used for low power ratings, e g in agricultural machinery, fans or saws. There is a risk of cracking, especially at the bottom of the groove, due to corrosion.

71

Figure 5-9. Removable hub pulley (bushing pulley). The bushing is equipped with a keyway so that it can be mounted on the shaft with the aid of keys. Furthermore, it is split axially so it can be clamped to the shaft. The clamping and locking onto the pulley is with the aid of screws threaded in holes between the bushing and the pulley

Figure 5-10. Pulley of aluminium with steel hub

Figure 5-11. Stamped steel plate pulley commonly used for low power ratings

Plastic pulleys: Many of the modern engineering plastics, for instance amide plastic (nylon) and acetal plastic, are well suited for pulleys for low power ratings. In small-scale production pulleys may be manufactured by machining, but for large runs injection moulding offers very competitive prices, figure 5-12. Other advantages are low weight, good corrosion resistance and good cleanliness due to the absence of corroded material. The operating temperature of plastic pulleys should not exceed 100°C. Another limitation

Figure 5-12. Pulley made from amide plastic (nylon)

Characteristics	Pulley type				
	Standard (cast iron/ steel)	Removable bushing (cast iron/ steel)	Aluminium	Stamped steel plate	Plastics
Cost	moderate	high	moderate	low	low
Ease of installation		+			
Ease of replacement		+			
Suitable for long manufacturing runs (high power ratings)	+				
Suitable for long manufacturing runs (low power ratings)			+	+	+
High power ratings	+	+			
Low weight			+	+	+
Corrosion resistance			+		+
Suitable for multi-groove design	+	+			

Figure 5-13. Some comparative data for V-belt pulleys of different materials and designs

compared to metallic materials is the normally larger tolerances of plastic products. Examples of application are for instance food industry, sewing machines and toys.

A summary and comparison of the various types of pulleys can be found in figure 5-13.

5.2.5 Balance

Good balancing of pulleys is necessary to avoid extra inertial loads and vibrations which may also give extra stress loadings on other parts of the belt drive, e g shafts or bearings. Specified limits of residual imbalance are for instance given in ISO 254. Requirements increase with increased face width and pulley speed. For pulleys with small face width and not too high speeds balancing in one plane (static balancing) is sufficient, while large face widths and high speeds, e g above 30 m/s, require balancing in two planes (dynamic balancing).

5.3 Pulleys for flat transmission belts

A simple pulley for a flat transmission belt needs only flat (cylindrical) pulley surface. To reduce the risk that the belt might run off the pulley, it can be flanged, which, however, is usually avoided as it can cause wear on the belt edges. Preferably the pulleys are made with a camber (crown), figure 5-14. The pitch (height of the arch) is usually around 0,3% of the pulley diameter and increases also at large diameters with the width of the pulley. It can be shown that a flat belt always strives to turn in the direction of its highest strain, i e run symmetrically with reference to the top of the crown.

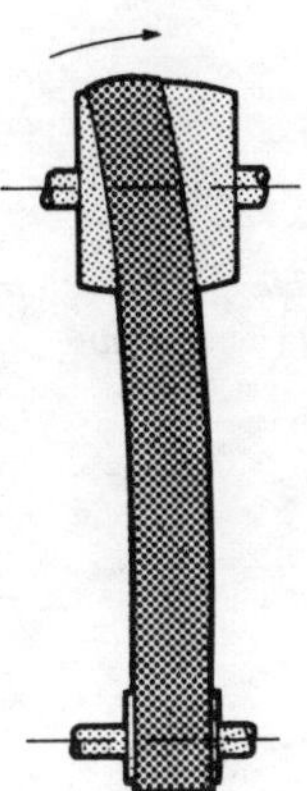

Figure 5-14. Flat belt drive with one pulley crowned

5.4 Pulleys for V-ribbed belts

Pulleys for V-ribbed belts may be constructed as pulleys for flat belts but are grooved to fit the ribs of the V-ribbed belt, figure 5-15. As a wedge effect is usually not strived for, the ribs of the belt can completely fill the grooves on the pulley. The groove and rib construction offers good side stability so that camber (crown) is not required.

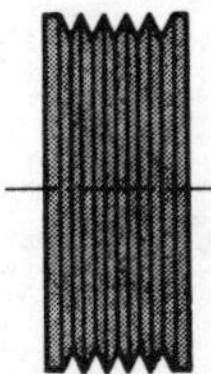

Figure 5-15. Pulley for V-ribbed belts

Figure 5-16. Flanged bushing pulley for synchronous belts

5.5 Pulleys for synchronous belts

Synchronous belts will require special pulleys with recesses to fit the size and pitch of the cogs, figure 5-16. Flanges are often used to prevent belts from running off the pulleys. Pulleys are manufactured with integral hubs as well as with bushings. The pulleys are fairly expensive which contributes to the higher price level for synchronous belt drives as compared to other belt drives. Besides pulleys of metallic materials plastic pulleys are often used, especially in large runs, made for instance from glass-fibre filled amide plastic and containing small amounts of graphite to reduce friction and wear.

5.6 Variator pulleys

The principal of variable speed changers (variators) is demonstrated in figure 2-2. The variator may also include one non-adjustable pulley of the type described in section 5.2. The adjustable pulley or pulleys in the variator are split in two halves in a plane in the middle of the pulley groove and the two halves are axially adjustable in relation to each other to permit the V-belts

76

A   B

Figure 5-17. Adjustable-speed drive pulley
A. Pulley halves close together, large pitch diameter
B. Pulley halves apart, small pitch diameter

to operate at a variable diameter. The pulley is then a component in a larger, often rather complicated mechanism. Even if some dimensions, angles and requirements on surface finish are standardized the final choice of materials and construction will be made to fit the special application in question, cf example in figure 5-17.

6. Materials

6.1 Introduction

In this chapter materials used in belts, especially V-belts, will be discussed. They can mostly be classified as either rubber materials or reinforcing materials. The development of improved materials during the last decades has been of prime importance for the development of new belts with improved power rating and longer life, figure 6-1.

6.2 Polymeric materials

With exception for some metallic or ceramic reinforcement materials, like steel-cord or glass-cord, belts are manufactured from and thus composed of various polymeric materials. Polymeric materials contain as the coherent and character-forming component an organic polymer. Organic polymers can be described as giant molecules, usually in the form of long flexible chains, and for the most containing carbon and hydrogen atoms and in certain cases also containing oxygen, nitrogen or chlorine atoms. Polymeric materials contain in addition, which is especially true for rubber materials, large amounts of additives like fillers, plasticizers, protective agents or agents intended to form chemical crosslinks during curing or vulcanization.

The classification of important groups of polymeric materials is given in figure 6-2. The decisive factor of classification is thus the degree of elastic extensibility of the material. Elastomers are polymeric materials which have an elastic extensibility, without rupture occurring, of at least 100%, i e that can be elongated to at least their double original length without rupture and after unloading elastically return to approximately their original undeformed dimensions. Polymeric materials with lower elastic extensibility, mostly only a few percent, and generally with higher stiffness than the elastomers are termed plastics. Polymeric materials with high strength and stiffness, obtained by orientation during stretching of thin filaments, are termed fibres.

Year	Example for belt drive with given power rating			Profile	Underside	With (0) or without (S) textile wrapping	Reinforcement	Rubber type: natural/styrene rubber (NR/SBR) or chloroprene rubber (CR)
	Number of belts	Relative, useful tension per belt	Belt speed, m/s					
1940	10	1	20	classical	uncogged	0	cotton cord	NR/SBR
1950	7	1,4	20	classical	uncogged	0	rayon cord	NR/SBR
1960	5	1,7	30	narrow	uncogged	0	polyester cord	CR
1970	4	1,9	35	narrow	uncogged	0	improved polyester cord	CR
1980	3	2,0	45	narrow	cogged	S	improved polyester cord	CR

Figure 6-1. Historical data showing how improvements in materials and construction for V-belts have made higher belt tension and belt speed possible, thus allowing a given power rating with a lesser number of V-belts of given belt width

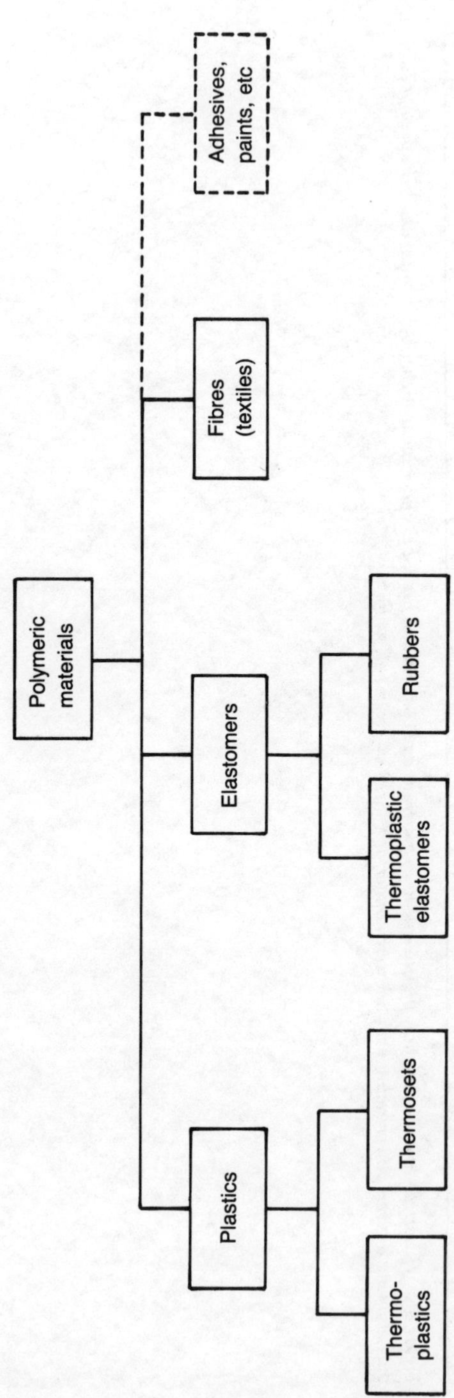

Figure 6-2. Classification of various types of polymeric materials

All polymeric materials are plastically deformable during at least one stage of their processing (shaping). One way of stabilizing the shape of the products after the plastic forming is by cooling. These materials can then be reshaped by heating them to their melt temperature. Plastics of this type are termed thermoplastics, corresponding elastomers are termed thermoplastic elastomers. The shape of the products can also be stabilized after plastic forming by introduction of chemical bonds between the polymer chain molecules, so-called crosslinks, which will fix their position. Elastomers that get their shape stabilized in this way are termed rubbers and the crosslinking process curing or vulcanization. Plastics, that are crosslinked, are termed thermosets and the crosslinking process either curing or hardening. Rubber is the dominant group among the elastomers, while thermoplastic elastomers still have only a limited market. Thermoplastics dominate the plastics group, but thermosets still have a considerable market share.

6.3 Components in a V-belt and their function

6.3.1 Components

The placement of the most important components in a V-belt is demonstrated in figure 6-3. As already mentioned the use of V-belts without textile wrapping, so-called cut V-belts, is now on the increase. Special varieties of V-belts may contain additional components, e g extra reinforcement layers to protect against wear, bending stresses, etc.

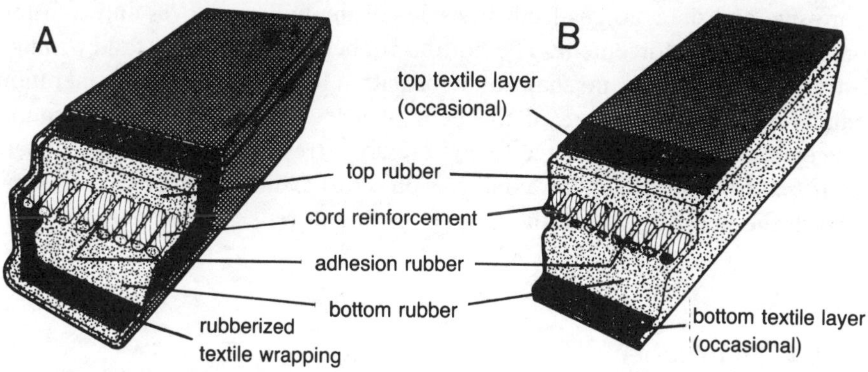

Figure 6-3. Components in a V-belt
A. With textile wrapping, longer belts (above 2,5 m in length) may contain several layers of reinforcement
B. Without textile wrapping (cut V-belt)

6.3.2 Reinforcement

The function of the reinforcement in the V-belt is to take up the tension forces in the belt which are necessary for power transmission and for pressing the belt against the pulleys in cases where power is transmitted by frictional forces. The reinforcement usually has a tensile modulus (Young's modulus, E-modulus) which is in the order of 1 000 times larger than that of the surrounding rubber material and a tensile strength in the order of 100 times larger. These figures demonstrate the dominant role the reinforcement plays in taking up tensile forces in the belt, where the rubber material, despite its much larger volyme compared to the reinforcement, only takes up a few percent of the tensile forces. Additional requirements on the reinforcement are good adhesion to surrounding rubber materials, good flex-fatigue resistance and good dimensional stability whilst under operating conditions. Dimensional changes during service, usually a gradual increase in length (creep), will require continued maintenance and adjustment of belt tension. High-modulus reinforcement materials give small deformations when in service, of great advantage in synchronous belts. Low-modulus reinforcement materials with high elastic extensibility can absorb large shock loads and are preferred where this property is important.

6.3.3 Bottom rubber

The rubber material that fills out the trapezoidal section below the reinforcement, is termed bottom rubber or base rubber. It must withstand the compressive and shear forces from the sides of the pulley grooves and transmit them to the reinforcement. The bottom rubber must also withstand bending stresses and have low mechanical deformation losses to keep heat generation and working temperature of the belts at a low level. Cut V-belts require exceptionally good resistance to compressive stresses in the bottom rubber. The material must have low compression set to avoid plastic deformation by which the belt will sink in the groove leading to reduced belt tension.

6.3.4 Top rubber

The rubber material that fills out the V-belt section above the reinforcement is termed top rubber. It forms a relatively thin layer from which it is required that it shall withstand repeated bending of the belts without fatigue cracking.

6.3.5 Adhesion rubber

Adhesion rubber or cushion rubber is the rubber that fills out the space between the bottom rubber and the reinforcement, the space between reinforcement cords and the space between the reinforcement and the top rubber. The adhesion rubber shall have good adhesion to all the other components and is usually fairly soft to flow well during manufacture and thus fill out all the interstices between the other components.

6.3.6 Textile wrapping

The rubber-impregnated and rubber-coated fabric forms the outer layer of wrapped V-belts. The prime function of the rubber-impregnated textile wrapping is to give good friction and good wear resistance in the contact between V-belt and pulley groves. The textile wrapping further contributes to the cohesion between the components in the V-belt under the influence of the large compressive and shear forces from the sides of the pulley grooves, and in general increases the stability of the V-belt. A drawback is that the bending resistance of the V-belt will be increased and this limits its ability to run over small diameter pulleys. In cut V-belts the absence of textile wrapping is compensated for by use of bottom rubber with high friction, good wear resistance and sufficient stiffness to withstand large compressive forces.

6.4 Rubber materials

6.4.1 Important properties

Rubber materials can be characterized by a number of properties. The most important ones for V-belts are briefly described below with reference to relevant standards, primarily ISO standards:

Hardness is determined as the resistance to elastic penetration of a needle pressed against the rubber surface (ISO 48). Hardness varies between 0 (no penetration resistance, infinitely soft) and 100 (no penetration, infinitely hard) IRHD (International Rubber Hardness Degrees). Most rubber materials have a hardness in the range 50–80 IRHD. Hardness can be taken as a measure of the stiffness (Young's modulus) of the material and can be roughly correlated with the modulus.

Tensile strength is measured at break (ISO 37) and values are given in MPa. Most rubber materials used in V-belts have a tensile strength in the range of

10–20 MPa. Laboratory figures usually refer to ideal conditions, i e thin test specimens, short loading times, room temperature, and freshly manufactured material. Lower values are to be expected for materials in V-belts in service due to heavier material sections, longer times of loading, elevated temperatures, imperfections that form starting points for cracks, etc. Engineering calculations must therefore apply a safety factor of at least 10 in relation to laboratory tensile strength obtained under ideal conditions.

Elongation at break is usually given in percent of original, undeformed length (ISO 37). Typical values are normally in the order of several hundred percent which offers a good safety margin against the strain in a V-belt during operation, e g during bending over the pulleys.

Tear resistance (ISO 34) is the force per unit width (N/m) needed to propagate a crack in a defined test specimen. Values exceeding 10 kN/m can be considered as excellent and values in the range 5–10 kN/m as good. Good tear resistance is desirable in components exposed to local, high stress concentrations in combination with surface imperfections.

Coefficient of friction is an important property for rubber materials, in contact with pulley groove sides which shall transmit high frictional forces. Measured values depend strongly on conditions of measurement such as contact pressure, contact surface area, speed of sliding, surface conditions, etc. Generally valid test methods are not available. It is thus recommended to make measurements under conditions as close as possible to the service conditions. The determination of frictional properties of materials used in V-belts should therefore preferably be made by manufacturing V-belts from the materials in question and testing them on a belt drive, for instance by determining the power that can be transmitted under varying conditions as a function of slip. Rubber materials of high hardness have generally a lower coefficient of friction than soft ones. Water is a good lubricant for rubber and reduces the coefficient of friction considerably. The coefficient of friction of rubber materials usually lies within the range 0,3–1,5 and is reduced to approximately half the original value by water lubrication. For rubber products under specified service conditions a narrower interval of coefficient of friction can be given. Thus rubber tyres on dry roads usually have a coefficient of friction of 0,6–0,8, flat transmission belts 0,3 and V-belts in a V-shaped groove 0,5 (apparent coefficient of friction), cf figure 12-8.

Wear resistance is also a property for which relevant data can only be obtained by field testing under service conditions. Standardized laboratory test methods can be found but give poor correlation with practice. Under certain wear conditions, e g in rolling contact between a tyre and road surface or for handling of particulate materials like gravel, ore, cement, etc, the wear resistance of rubber is superior to all other materials. The wear on rubber belts in transmission drives is usually very small, but can become catastrophic under abnormal conditions, for instance during high slippage.

Plastic flow under long lasting tension loading can lead to a gradual increase in length (creep), to a permanent deformation after unloading (tension set/ISO 2285), to a permanent deformation after compressive loading (compression set/ISO 815) or to a decline in force at constant deformation (stress relaxation/ISO 3384). Plastic flow will increase with increase in temperature and time. As it is the reinforcement that takes up most of the forces in a belt, the plastic flow properties of the rubber do not play the same dominant role as in many other rubber products.

Cold stiffening will be observed with all rubber materials if the temperature is low enough. The rubber material eventually becomes so hard and brittle that it will break during bending. Cold resistance can be given as the lowest temperature at which the rubber can be bent without breaking (brittle-point/ ISO 812), as the gradual stiffening with decrease in temperature (ISO 1432) or as the elastic recovery of stretched and deep-frozen specimens as a function of temperature (ISO 2921). For not too large or rapid deformations most rubber materials have a cold resistance of at least −40°C. By choice of special rubber types cold resistance may be extended down to −80°C. When starting up belt drives after long standstills at very low temperatures, e g below −30°C, care must be taken. Enclosing the belt drive to protect against extreme cold or warming the belts before start is recommended.

Ageing is the continuous and usually deleterious change in properties of rubber exposed to elevated temperatures for long times. Ageing is a chemical and irreversible process, usually an oxidation in the presence of atmospheric oxygen. It is usually accompanied by an increase in hardness and reduction in elongation at break, while tensile strength may change in either direction. Ageing is measured in the laboratory by following the changes in these properties after exposure to elevated temperatures (ISO 188). When the material becomes so hard and brittle that it is no more rubberlike it has reached the end of its useful life. If lifetime is plotted on logarithmic scale versus the inverted value of absolute temperature straight lines are obtained, figure 6-4. This is in agreement with Arrhenius law, which is as valid for the ageing process as for other chemical reactions. The slope of straight line curves in figure 6-4 corresponds to an old rule of thumb for the ageing of rubber, i e that lifetime will be approximately cut in half for a temperature increase

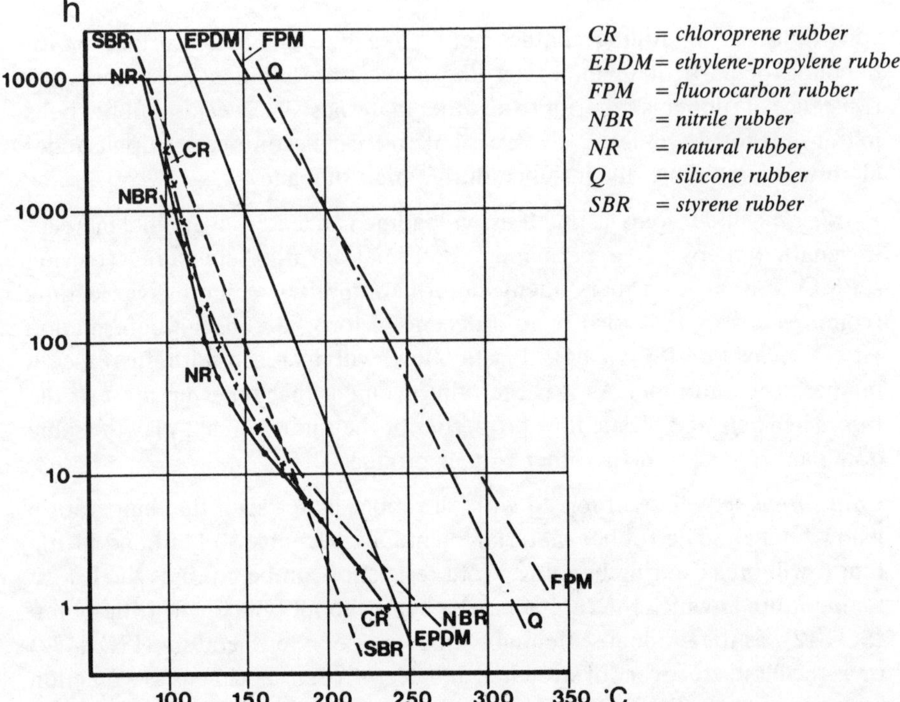

Figure 6-4. Examples of life expectancy curves for rubber materials based on different rubber types (values obtained at Trelleborg AB). Lines are drawn for life expectancy in h (logarithmic scale) versus ambient temperature (°C) which with fair approximation give straight lines. To correspond to the Arrhenius equation the temperature should preferably be plotted as the inverse of the absolute temperature which, however, will not markedly affect the shape of the curves within the temperature interval studied. Life expectancy is here defined as the time of ageing at a specified temperature after which the tensile strength of the material falls below 4 MPa, or alternatively the elongation at break falls below 100%.

of 10°C and roughly doubled for a temperature decrease of 10°C. Practical experience shows that this is not only valid for the rubber material itself but also for the working life of V-belts. This points to the importance of keeping belt operating temperatures as low as possible in order to achieve long working life. For thermoplastic materials including several reinforcing materials, an additional factor must be taken into account, viz the softening and increase in plastic flow with increase in temperature, whereby modulus and tensile strength are decreased and creep increases. When the softening temperature is approached, the physical and reversible softening process will become of greater importance for the products, than the irreversible chemical ageing processes.

Mechanical damping during deformation of the rubber material leads to a loss energy, that is transformed into heat and raises the operating temperature of the belt. The simpliest way of testing this property is by a rebound test (ISO 4662), in which the loss in rebound height can be taken as a measure of the mechanical damping. Values depend strongly on magnitude of deformation, speed or frequency of deformation and temperature. For more accurate estimations a test method should be used under conditions as close as possible to the actual operating conditions.

Ozone resistance is the ability of the material to withstand ozone without cracking at atmospheric concentrations (1–10 pphm) or in the neighbourhood of electrical machinery. The cracking will only appear when the rubber is under tension. Cracks are mostly small and shallow but may act as starting points for deeper fatigue cracks. The property is studied in the laboratory by exposing elongated test specimens to specified ozone concentrations (ISO 1431).

Fatigue resistance is the ability to withstand small, repeated stresses without damage or breakage. Damage can be either cracks starting from the surface of the product or, for composites, delaminations between components. Crack growth can be measured by laboratory testing (ISO 132/ISO 133). As fatigue phenomena depend in a complicated way on stresses and strains (amplitude/ frequency), surrounding atmosphere, temperature, etc, reliable results can only be obtained by fullscale testing of belts under normal or exaggerated operating conditions, cf chapter 15.

Oil resistance is the ability of the material to withstand exposure to oils, propellants, fuels, solvents, etc. These fluids have a tendency to be absorbed by the rubber material, whereby its volume changes and mechanical properties deteriorate. Oil resistance, also termed swelling resistance, is usually determined by measuring volume change of test specimens after immersion in liquids for standardized conditions of time and temperatue (ISO 1817).

Electrical properties often involve demands placed on resistivity, which should not be too high so as to avoid the build-up of electrostatic charges on the belt. Various laboratory testing methods are available but best results are obtained by measurement on a V-belt in a drive (ISO 1813). The volume resistivity of rubber can cover a very wide range, 10^{-2}–10^{15} $\Omega \cdot m$, and depends mainly on the type and amount of carbon black used as a filler.

Flammability can be an important property in certain belt drives. Several laboratory methods are available for comparing rubber materials in this respect, but also here tests on actual V-belts are recommended to get good agreement with service values, e g according to the British standard BS 3790.

6.4.2 Important rubber types

Only a few of the roughly 20 different existing types of rubber are used in V-belts, figure 6-5. In some, very special applications other types of rubber may also be used. Of the types listed in figure 6-5 NR/SBR and CR account for more than 95% of the volume of rubber in belts. The properties of the different types of rubber used in V-belts can be summarized as follows:

Natural rubber (NR) is obtained from the rubber tree, Hevea brasiliensis, mainly from plantations in South-East Asia. It is the first type of rubber to be used and until about 1930, also the only type available. Natural rubber is still one of the best rubber types in respect of mechanical properties such as tensile strength, wear resistance and low mechanical damping. It is further easy to process, especially when building composite products. Limitations are found in oil resistance, ozone resistance and heat resistance (ageing). The first flat belts and V-belts were manufactured with natural rubber and it is still widely used thanks to its good processibility, good mechanical properties and comparatively low price.

Styrene rubber (SBR) is today the rubber type with the largest production volume and is a synthetic material manufactured from oil. Differences in properties compared to natural rubber are marginal, involving mainly a somewhat better heat-ageing resistance. The price is also comparatively low.

Chloroprene rubber (CR) is likewise a synthetic rubber, used since 1930. Well-known tradenames are Neoprene and Baypren. Compared to NR or SBR it has better heat-ageing resistance, better ozone resistance and fair oil resistance to splashes or other accidental contact with oil. Mechanical and fatigue properties are good. Chloroprene rubber has therefore become the dominant rubber type for use in high-quality V-belts.

Nitrile rubber (NBR) has better oil resistance than CR, about the same heat-ageing resistance, but not as good ozone resistance. It is primarily used in belts heavily exposed to oil.

Urethane rubber (AU/EU) is processed by special techniques (mainly by casting of liquid components) and is characterized by very good mechanical properties, excellent oil resistance and excellent ozone resistance. Its limitations are mainly the high price, the rapid stiffening at low temperatures and poor hydrolytic resistance under the influence of acids, hot water or hot, humid atmospheres. With special compounding, materials with very high coefficients of friction can be obtained. Urethane rubber is therefore mainly used in wide-angle V-belts.

Type	Chemical composition	Mechanical properties (high quality materials)			Working temperature, °C		Resistance to		
		Hardness IRHD	Tensile strength MPa	Wear resistance	Max	Min	Ozone (outdoor)	Hydrolysis (steam, acids)	Hydrocarbons (gasoline, oil)
Natural rubber (NR)	cis-polyisoprene	40–80	15–30	1	100	−70	3	1	4
Styrene rubber (SBR)	copolymer of normally 20–30% styrene and 70–80% butadiene	50–80	10–25	1	120	−60	3	1	4
Chloroprene rubber (CR)	polychloroprene	40–80	10–25	2	130	−50	2	1	3
Nitrile rubber (NBR)	copolymer of 20–45% acrylonitrile and 55–80% butadiene	55–85	10–25	2	130	−60	3	2	2
Urethane rubber (U)	polyesters (AU) or polyethers (EU) reacted with polyisocyanates	70–95	30–50	1	120	−40	1	4	2

Figure 6-5. Comparative data for rubber types commonly used in V-belts. The grading of resistance implies: 1=excellent, 2=good, 3=fair and 4=poor

Thermoplastic elastomers have generally higher stiffness than most rubber materials and can therefore be used in V-belts without the need of reinforcing materials. One such example is V-belts made from polyester-polyether thermoplastic elastomers (Hytrel). They are usually manufactured in long lengths with open ends and then heat-welded to endless belts with the aid of special welding fixtures, cf figure 9-12. Compared to normal, reinforced V-belts they have low power rating and wear out fast. Their use is therefore limited to temporary applications, small power ratings and where installation of endless belts may be difficult.

6.4.3 Rubber compounds

A rubber material contains, in addition to the rubber polymer, a number of additives, the most important ones being:

Fillers are solid particulate materials which increase the hardness and stiffness of the compound. They are often used at high loadings. Fillers of extremely low particle size (particle diameter below 50 nm) are called reinforcing fillers as they improve tensile strength, tear and wear resistance of the rubber material. Such finely divided materials can only be produced economically by gas-phase reactions. The dominant reinforcing filler used in rubber is therefore carbon black (soot), obtained from the incomplete combustion of oil with a deficit of oxygen. The wide use of carbon black explains why most rubber materials are black in colour. Examples of non-reinforcing fillers with larger particle size can be found among natural, mineral products such as clay or whiting.

Plasticizers, also called softeners, are liquid, semi-solid or resinous products, which reduce hardness and stiffness and facilitate processing. Used in large amounts they will also reduce tensile strength and wear resistance. Examples of plasticizers are hydrocarbon oils, fats, fatty oils, esters, etc.

The curing system includes all ingredients necessary to bring about the vulcanization (curing) reaction. Components include *curing agents,* usually sulphur, *accelerators,* usually complicated organic compounds with the purpose of speeding up the vulcanization reaction, *retarders* with the purpose of preventing the vulcanization reaction from starting at processing temperatures and *activators,* usually zinc oxide and stearic acid, which facilitate the action of the accelerators. The complete vulcanization system usually represents less than 5% of the weight of the compound. Often a combination of so-called primary and secondary accelerators is used. CR requires other curing systems than NR/SBR, so that for instance zinc oxide would be used as a curing agent instead of sulphur.

Ingredients	Parts by weight	Function
Natural rubber SMR 20	100	Rubber polymer
Carbon black N 774 (SRF)	60	Filler
Naphthenic oil	10	Plasticizer
Zinc oxide	5	Activators
Stearic acid	1	
Sulphur	2,5	Curing agent
Sulphenamide	0,8	Accelerators
Thiuramdisulphide	0,4	
Trimethyldihydroquinoline	1,5	Antioxidant
Paraphenylenediamine	0,5	Antiozonants
Microcrystalline wax	1,5	
Total 183,2		

Figure 6-6. Example of a rubber mix recipe indicating the function of the various ingredients used

Protective agents can be *antioxidants,* complicated organic compounds with the purpose of slowing down the ageing reaction, *antiozonants,* with the purpose of preventing ozone cracking or *flame retardants,* added to reduce the flammability of the material.

In figure 6-6 is given an example of a recipe for a rubber compound, giving the function of the various ingredients.

6.4.4 Choice of rubber materials for various components

The exact composition of the rubber materials used in various components of a V-belt is a well-guarded secret with the manufacturer. General principles that guide the choice of materials are exemplified below:

Bottom rubber can either be manufactured from NR/SBR or CR, cf example in figure 6-7. CR is preferred in quality belts, especially in cut V-belts, which have no rubber-coated wrapping to protect the surface from attack by oxygen, ozone or oil. CR can also better withstand the high working temperatures accompanying increase in belt speed or power rating. To increase the resistance of the bottom rubber against the high compressive forces from the sides of the pulley grooves, it is usually stiffened by admixture with short fibres. During manufacture the fibres are oriented transverse to the length direction of the belt, whereby maximum resistance towards compressive forces is obtained with only a moderate increase in the bending resistance. By applying this technique the stiffness in the transversal direction can be 3–5 times higher than in the longitudinal direction.

Application	High quality bottom rubber for cut V-belts	Medium quality bottom rubber for belts with textile wrapping	High quality adhesion rubber	Coating rubber on textile wrapping
Ingredients				
Chloroprene rubber	100	--	100	100
Styrene rubber	--	100	--	--
Conductive carbon black N472 (XCF)	--	--	--	30
Carbon black N650 (GPF)	35	--	40	--
Carbon black N774 (SRF)	--	100	--	--
Silica	--	--	30	--
Cotton flock, 4–6 mm	20	--	--	--
Hydrocarbon oil	5	5	8	20
Rosin	--	5	--	--
Zinc oxide	8	3	5	5
Magnesium oxide	4	--	4	4
Stearic acid	2	2	2	1
Special accelerator for CR	--	--	0,25	--
Primary accelerator for SBR	--	2	--	--
Secondary accelerator for SBR	--	0,2	--	--
Sulphur	--	0,4	--	--
Antioxidant (mixture of alkylated diphenylamines and paraphenylene-diamines)	3	1	3	3
Total (parts by weight)	177	218,6	192,25	163
Properties				
Hardness, IRHD	80	78	70	60
Tensile strength, MPa	13	18	14	15
Elongation at break, %	150	210	460	500

Figure 6-7. Examples of recipes for rubber materials to be used in various components of V-belts

Top rubber is usually of the same type as the bottom rubber, i e based either on NR/SBR or CR, but without short fibres which would have no function here.

Adhesion rubber is usually based on the same rubber types as the bottom rubber and top rubber, but is generally softer to flow better and fill out interstices better, cf example in figure 6-7. It is of great importance that the adhesion rubber adheres well to the reinforcement as well as to surrounding rubber materials.

Coating rubber (friction) on the textile wrapping is usually based on CR to offer good resistance to attack by ozone, oxygen or oils.

6.5 Reinforcement

6.5.1 Important properties

In evaluating and comparing various reinforcement materials the following properties are the most important ones:

Density is important as for instance tensile strength and stiffness (modulus) are not only given as force per unit area but also as force per unit area and unit mass (weight). The difference in density between organic fibres and inorganic ones is considerable.

Stiffness can be given either as Young's modulus (E-modulus) or as the stress needed to reach a certain strain, e g 2%. To include density the unit tex is used, defined as the mass in grams of 1 km of the thread. Modulus is then given in mN/tex. By multiplying modulus in mN/tex by density in Mg/m^3, E-modulus units in MPa will be obtained.

Tensile strength can also be given either in mN/tex or in MPa.

Elongation at break is given as percent of original length.

Shock absorption ability can be estimated from the diagram area below the stress-strain curve up to the breaking point, cf figure 6-9. Good shock absorption is usually to be found with fibre materials with high elongation at break and usually not a too high tensile strength or modulus, while high-strength and high-modulus fibre materials usually have only modest shock absorption ability.

Flexural fatigue resistance is also related to the stiffness of the fibre material. Best fatigue properties are obtained with fibre materials of a not too high stiffness. A number of laboratory test methods are available for evaluating this property, but most reliable results are obtained when testing under conditions as close as possible to the practical case. Fatigue properties of various reinforcing materials for V-belts are thus best evaluated by testing on V-belts under normal or somewhat exaggerated operating conditions, cf section 8.3.

Heat generation of the reinforcement should be as low as possible to keep up mechanical efficiency and keep down temperature rise during operation, caused by heat generation. As simple laboratory tests are not available the property has to be studied on V-belts under operating conditions.

Moisture pick-up can be measured as moisture content or moisture regain. Natural fibres usually have high moisture pick-up, meaning that their properties and dimensions vary under the influence of varying ambient humidity. Modern synthetic fibres are not so sensitive, which reduces the importance of this property.

Maximum allowable temperature is limited by either heat ageing or thermal softening leading to decreased stiffness and strength and increased creep rate and shrinkage. Time, temperature and level of loading will influence the results obtained. Laboratory test methods are available in which these factors can be varied in different ways.

Creep will increase with time, temperature and stress. Several laboratory test methods are available for evaluation.

Shrinkage can occur at elevated temperatures when stresses in the fibre material that have been frozen-in during manufacture are released as a result of the so-called "elastic memory" of polymeric materials, cf section 6.5.4. The shrinkage observed will be a function of temperature, time and degree of frozen-in strain in the fibre material. Various laboratory test methods are available where these factors can be varied.

6.5.2 Fibre types

A summary of the various fibre types used in V-belts is shown in figure 6-8.

Cotton is the oldest fibre material used but is today only to be found as fabric reinforcement in flat transmission belts or as fabric wrapping on V-belts. In other areas cotton has been replaced by modern synthetic fibres, which have better mechanical properties, are not as sensitive to humidity and can today also compete pricewise.

Rayon first replaced cotton as reinforcement in V-belts but has now been largely replaced by modern synthetic fibres due to their improved mechanical properties, lower sensitivity to humidity and more favourable price.

Polyamide (nylon) fibre has a high energy and shock absorption ability and is for this reason used in some types of V-belts. Its high elastic extensibility is accompanied by lower stiffness (modulus) leading to increased deformation and reduced dimensional stability. Polyamide fibre is often a component in fabric used for the wrapping of V-belts and it is the standard material in the fabric used as surfacing on cogs in synchronous belts.

Polyester fibre is regarded as the best reinforcing material for transmission belts and has today probably more than 90% of this market. Its high strength,

94

Fibre type	Density Mg/m^3	Tensile strength		Modulus of elasticity		Elongation at break %	Impact resistance (shock absorption capability)	Flex fatigue life resistance	Useful long time up to °C	Heat shrinkage	Moisture resistance	Creep tendency at elevated temperatures
		per unit surface area MPa	per unit surface area and unit weight mN/tex	per unit surface area MPa	per unit surface area and unit weight mN/tex							
Natural fibres												
Cotton	1,54	550	350	4000	2500	10	fair	fair	130	low	fair	moderate
Regenerated fibres												
Rayon	1,52	750	500	8000	5500	15	good	good	100	low	poor	moderate
Synthetic fibres												
Polyamide (nylon)	1,14	1000	850	4000	3500	20	excellent	excellent	120	moderate	good	high
Polyester	1,38	1100	800	8000	8000	15	good	good	150	moderate	excellent	moderate
Aramid	1,44	2500	1800	40000	30000	5	good	good	220	none	excellent	none
Inorganic fibres												
Glass	2,54	2000	800	55000	20000	5	fair	poor	>500	none	good	none
Steel	7,8	2750	350	150000	20000	2	fair	poor	>500	none	fair (rust)	none

Figure 6-8. Comparative data for important fibre materials used as reinforcement in belts. Tensile strength and stiffness (modulus) with respect to the weight of the material (mN/tex) is obtained by dividing the normal figures with respect to area only (MPa) with density (Mg/m^3). Tex gives the weight in grams of 1 km of the textile thread. Figures given are for textile constructions normally used as reinforcement in belts

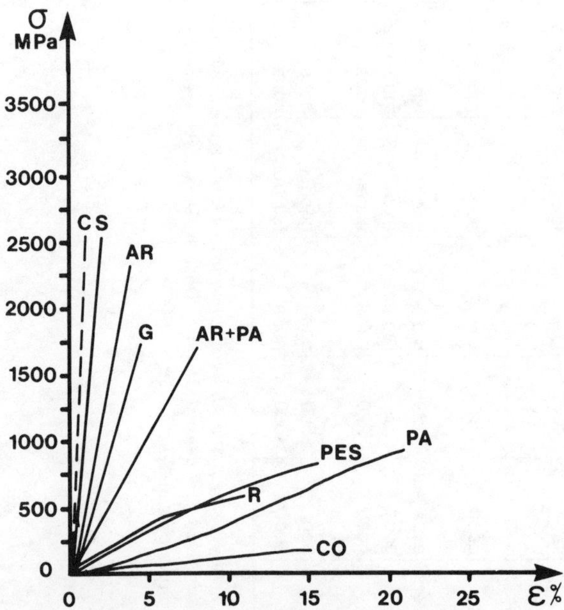

Figure 6-9. Stress-strain curves for some fibre materials used as reinforcement in V-belts (values obtained for treated yarn or cord at Trelleborg AB)
C=carbon, S=steel, AR=aramid, G=glass, PES=polyester, R=rayon, PA=polyamide, CO= cotton

stiffness and dimensional stability, in combination with good fatigue properties, are the most important factors.

High-modulus fibres include *aramid, glass* and *steel*. Their high stiffness and dimensional stability make them specially suited for applications, where this property is of importance, e g in synchronous belts.

A comparison of tensile stiffness of various fibre types is given by the stress-strain curves in figure 6-9. Increased stiffness generally leads to lower energy and shock absorption ability and to lower fatigue resistance. The low extension needed during installation can be both an advantage, as smaller allowances on shaft centre distance correction may be used, or a disadvantage, as installation will require smaller tolerances.

6.5.3 Textile constructions

Typical for textiles, e g used as reinforcement, is that their properties are not only determined by the fibre type, but also by the textile construction used. The basic element in the textile construction is the fibre, which can be

96

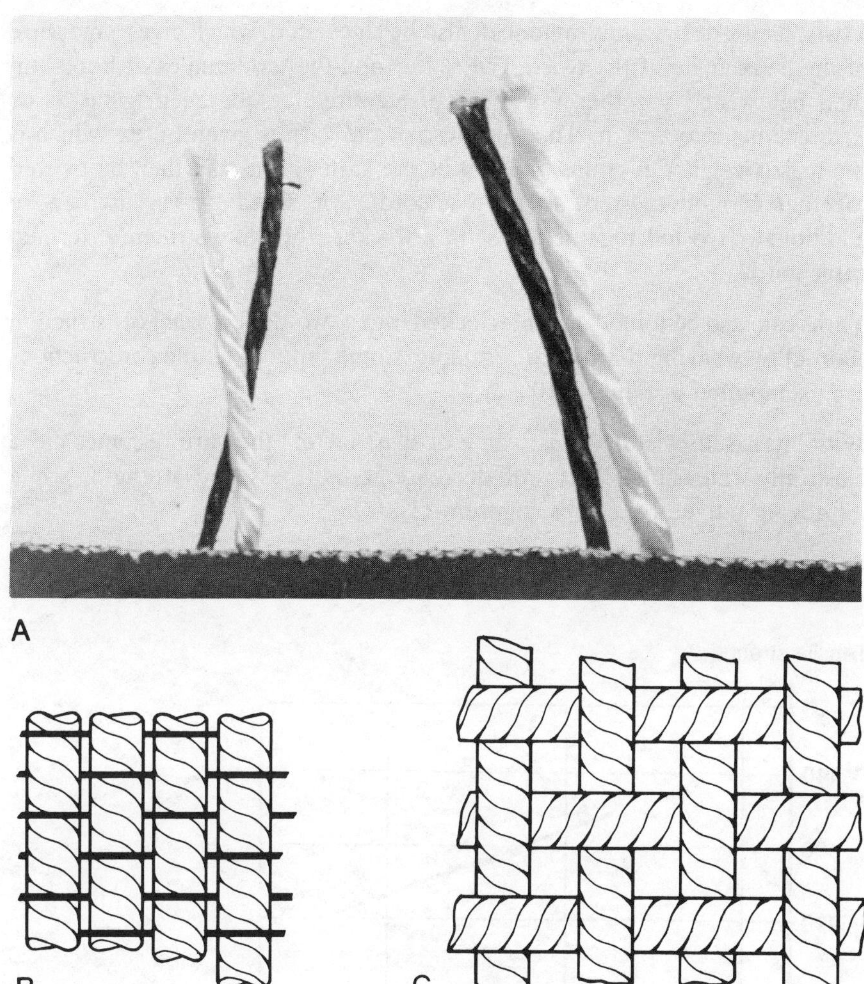

Figure 6-10. Examples of textile constructions used in belts
A. Cord (yarn)
B. Cord fabric
C. Plain weave

described as a thin thread with a diameter of 10–30 μm. These are brought together in bundles, which may or may not be twisted. The twist can be in either of two directions, usually designated as S- or Z-direction. The degree of twist can be given as the twist value, usually measured as the number of twists per metre, e g 400 m^{-1}. Taking the thickness of the threads into account a twist factor or twist multiplier can also be calculated, which gives a measure of the helix angle of the twist. Twisted or non-twisted bundles of fibres can then be twisted together in the reversed direction to the original S- or Z-directions into a yarn. The thickness of the yarn is given in tex, which is the mass (weight) in grams of 1 km of the yarn. Yarns can then be twisted together (double-twisted) to form a cord yarn. Cord yarns can then be additionally twisted together to form a thicker, rope-like structure, termed cable cord.

Yarns can also be joined and interlocked into a two-dimensional construction (fabric) by weaving, knitting or braiding. Some types of textile constructions are exemplified in figure 6-10.

With increased twisting (twist value or twist factor) the yarn becomes more elastically extensible. This will decrease its stiffness and strength, while improving fatigue resistance, figure 6-11.

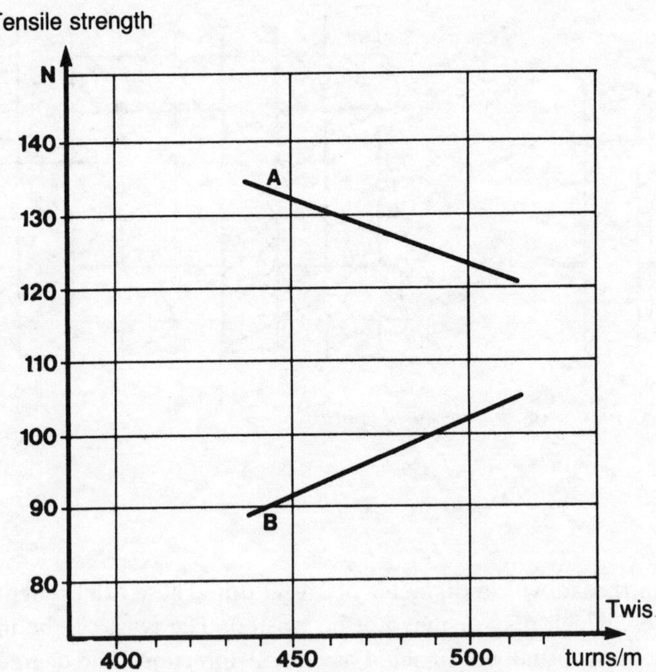

Figure 6-11. Effect of twist of yarn on tensile strength of the thread (A) before fatigue test, (B) after fatigue test (fatigue tested in rotor test, 200 min, at Trelleborg AB)

98

The design of a cord yarn construction is usually designated by a series of numbers, separated by multiplication signs. The first figure gives the thickness of the primary bundle of fibres in dtex and the following numbers indicate the number of bundles or yarns in the consequent twisting operations. Thus the designation 1 100×2×5 for polyester cord means that the operation starts with bundles with a thickness of 1 100 dtex of which first 2 bundles are twisted together and later 5 of these yarns additionally twisted together to form the final cord yarn. Aramid cord designated 1 670×1×2×5 indicates that the operation starts with bundles of 1 670 dtex, which are first twisted individually, 2 of the yarns so obtained are twisted together and finally 5 of the twisted yarns are in turn twisted together to form the final cord yarn.

Cord yarn and cable cord yarn are the most important types of textile constructions used as reinforcement in V-belts. Cord fabric may be preferred, especially in long belts. In the cord fabric the reinforcing cords in the warp direction are kept together by thin, loosely placed yarns in the weft direction. As stresses on the belts mostly act in their length direction, the best utilization of the reinforcement is obtained by having as much as possible of the reinforcement placed in the warp direction. One exception is the open-ended belts, joined by mechanical belt fasteners, e g certain types of flat transmission belts, where a considerable amount of reinforcement in the weft direction is desired to secure the fasteners in place. For textile wrapping on V-belts or on the cogs of synchronous belts fabrics with plain weave are mostly preferred. This fabric is bias-cut before it is wrapped around the belt core, so that the threads in the warp and weft lie at 45° angles to the length direction of the belt, whereby the bending stiffness can be kept as low as possible.

The blending of various fibre materials in the same textile construction offers additional opportunities. Thus bundles of various fibre materials can be twisted together to form yarns and cords. In fabrics different fibre materials may be used in the warp and weft directions to obtain different deformation properties in each of these directions. One example of such blended or hybrid materials is wrapping fabrics made from the blend of cotton and polyamide or polyester fibres which offers better strength and wear resistance than the pure cotton. Another example is to be found in cord yarns manufactured as a blend of aramid and polyamide or polyester fibre in order to get improved shock absorption ability and flex fatigue resistance, while, as far as possible, retaining the strength and stiffness of the pure aramid fibre.

Examples of some typical textile constructions for use in V-belts are presented in figure 6-12.

Application	Textile construction	Fibre type	Yarn/cord designation
Wrapping	plain weave	cotton	tex 40×2
Reinforcement (short belts)	cord yarn	polyester	dtex 1100×3×3
Reinforcement (long belts)	cord fabric	polyester	dtex 1100×2
Stiffener at bottom	plain weave	cotton	tex 74
Stiffener (variator belts)	cord fabric	cotton	tex 46×3×3

Figure 6-12. Data for some textile constructions used in V-belts. Cotton fabric was earlier used for stiffening but is now often replaced by short fibres mixed into the rubber

6.5.4 Viscoelastic properties

Most textile materials are polymeric materials exerting viscoelastic behaviour, i e they show both plastic deformation (creep) and elastic deformation. The plastic or viscous part of the deformation increases with time, temperature and stress level, while short-time loading at low temperatures mostly results in an almost pure elastic deformation. During plastic deformation the long polymer molecules become oriented in the direction of straining. By rapid cooling and freezing-in the molecules in their oriented state, a material with much higher strength and stiffness than the un-oriented material is obtained. When the oriented material is heated, the molecules will tend to return to their un-oriented, undeformed state – an effect known as "elastic memory". This will lead to shrinkage in the fibre. Exposing textile materials to elevated temperatures may thus result in either a permanent increase in length (creep) or a permanent decrease in length (shrinkage). Which effect that will dominate will depend on time, temperature and stress level, but will also depend on the conditions applied during manufacture and orientation of the fibre material. Individual fibres are, to a varying degree, stretched and oriented during their manufacture. To make optimum use of the textile, the textile construction should be stretched and oriented before usage, and this is usually carried out in connection with the impregnation of the textile to improve its adhesion to rubber, cf section 7.3. How the most important process variables during this so-called hot-stretch procedure influence the properties of the textile is exemplified in figure 6-13.

Property	If temperature is increased	If exposure time is increased	If tension is increased
Tensile strength	decreases	decreases	increases
Modulus of elasticity (stiffness)	increases	increases	increases
Shrinkage	decreases	decreases	increases
Creep	decreases	decreases	decreases
Flex fatigue life	increases	increases	decreases

Figure 6-13. Effect of the increase in tension, temperature and time in the hot-stretch procedure on important properties of the textiles treated

The most negative effect of the viscous and viscoelastic behaviour of the textile is the creep, i e the permanent increase in length of a belt during operation, figure 6-14. In favourable cases this may be compensated for by simultaneous shrinkage. Excessive slip of the belt caused by the creep can lead to temperature increases, which in turn lead to a shrinkage that automatically compensates for the creep. The increase in belt length due to wear may also be compensated for in the same way.

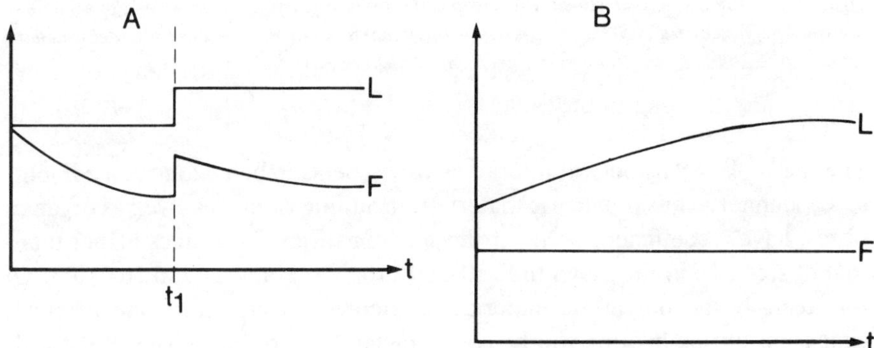

Figure 6-14. Principal changes in belt force F and belt length L with time of service for V-belts
A. Pre-tensioning with constant elongation (t_1 denotes time for re-tensioning)
B. Pre-tensioning with constant force

Dynamic loading of a V-belt in operation will not only lead to creep, but also to orientation of the material, by which its stiffness increases, figure 6-15. Creep starts at a high rate, but levels out and will gradually stabilize the length at a certain level.

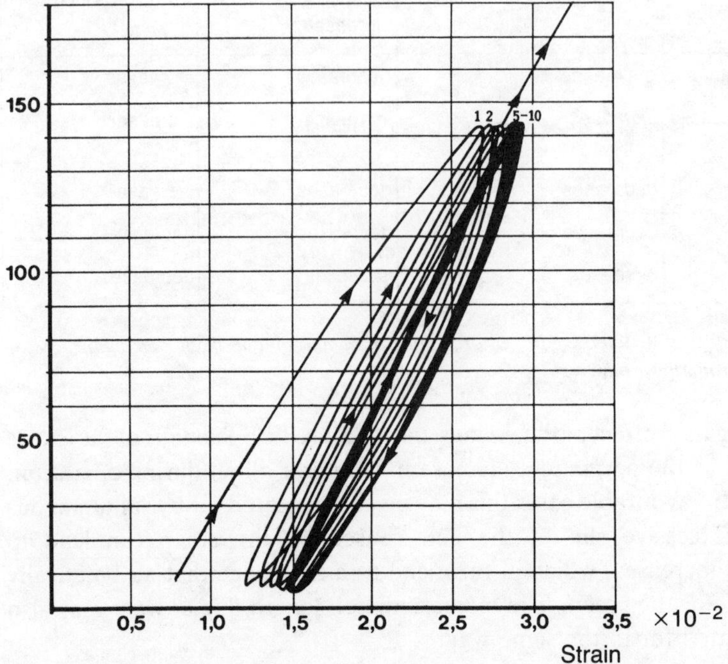

Figure 6-15. Stress-strain curves showing relaxation and creep effects by repeated deformations of treated polyester cord 1100×2×5 (figures on top of the curves denote the number of deformation cycles) up to 20% of fracture load (values from Trelleborg AB)

The thermal expansion will interact with viscoelastic behaviour in a complicated manner. Thus polymeric materials, including rubbers as well as organic fibres, have a coefficient of linear thermal expansion 5–20 times higher than that of steel, i e in the magnitude 50–$200\times10^{-6}/°C$ compared to $10\times10^{-6}/°C$ for steel. If the polymeric material is oriented as in a fibre the thermal expansion during heating will be counteracted by shrinkage. This effect will be amplified by the fact that an oriented polymer in a heated condition will have greater molecular mobility. Furthermore it will tend to increase its modulus in proportion to the absolute temperature and so it will strive to contract when kept at a given stress. The final result of all this will be that polyamide and polyester fibres show negative thermal expansion when heated, i e they will shrink, which normally leads to increased belt tension when the belt warms up during operation. Belts reinforced with glass fibres (coefficient of linear thermal expansion $7\times10^{-6}/°C$) will expand less than steel

102

when heated, which means that belts, e g on an automotive engine, tend to increase in belt tension when the engine and the belt reach high temperatures in operation.

6.5.5 Oriented plastic strip reinforcement

The high strength and stiffness of textile materials is a result of the molecular orientation imparted to the material during the spinning and consequent textile treatment operations. The same technique can be used to increase strength and stiffness of the material in the form of plastic film or plastic strip, figure 6-16. These oriented products can then be used as reinforcement in flat transmission belts with increased power ratings compared to older types of flat transmission belts with fabric reinforcement. The belts can be manufactured either endless or with open ends and later joined with the aid of chemicals. With plastic oriented strip reinforcement more reinforcement can be placed in a given volume and joining is easier.

Shape	Degree of orientation	Modulus of elasticity, MPa	Tensile strength, MPa	Elongation at break, %
Moulded articles	none	1000	80	200
Oriented strip	moderate	2000	200	50
Fibre	high	4000	1000	20

Figure 6-16. Tensile stress, elongation at break and modulus of elasticity for polyamide 6 at various levels of pre-stretching and orientation

7. Manufacture

7.1 Introduction

In this chapter the manufacture and course of operations for the production of V-belts will be discussed briefly. Other types of belts are manufactured with roughly the same technique, but differences will not be discussed here.

Figure 7-1 shows a principle flowsheet for the manufacture of V-belts. The final steps will differ depending on whether V-belts with textile wrapping or cut V-belts are to be manufactured. To obtain end products of high and even quality, high precision and extensive quality control must be applied in all manufacturing steps. In the first steps of the production materials of even quality must be used. In the following steps – primarily building and curing operations – the placement of the components must be precise within narrow tolerances and not be disturbed, and the processes carried out so that good adhesion is secured.

7.2 Mixing

The first step in rubber processing is mixing, i e to mix the rubber polymer with all the additives that are discussed in section 6.4.3. Mixing is usually carried out batchwise in batches of 50–500 kg, but continuous mixers are also available. Mixing was earlier performed on open roll mills, but today it is carried out in closed internal mixers, so-called Banbury mixers. The rubber mass is here kneaded and sheared in a closed chamber by two counter-rotating rotors equipped with 2–4 wings, while being kept under pressure from a ram driven by an air piston in the feed hopper, figure 7-2. The high viscosity of the material requires large forces and high energy input for the mixing. The mixing energy is transformed into heat in the material which is then warmed up. During the short mixing cycles, normally of a few minutes, the heat generated cannot be completely dissipated by convection to the cooled walls and rotors of the mixer. Mixing is therefore usually performed in two steps. In the first step all ingredients, with the exception of the curing system, are

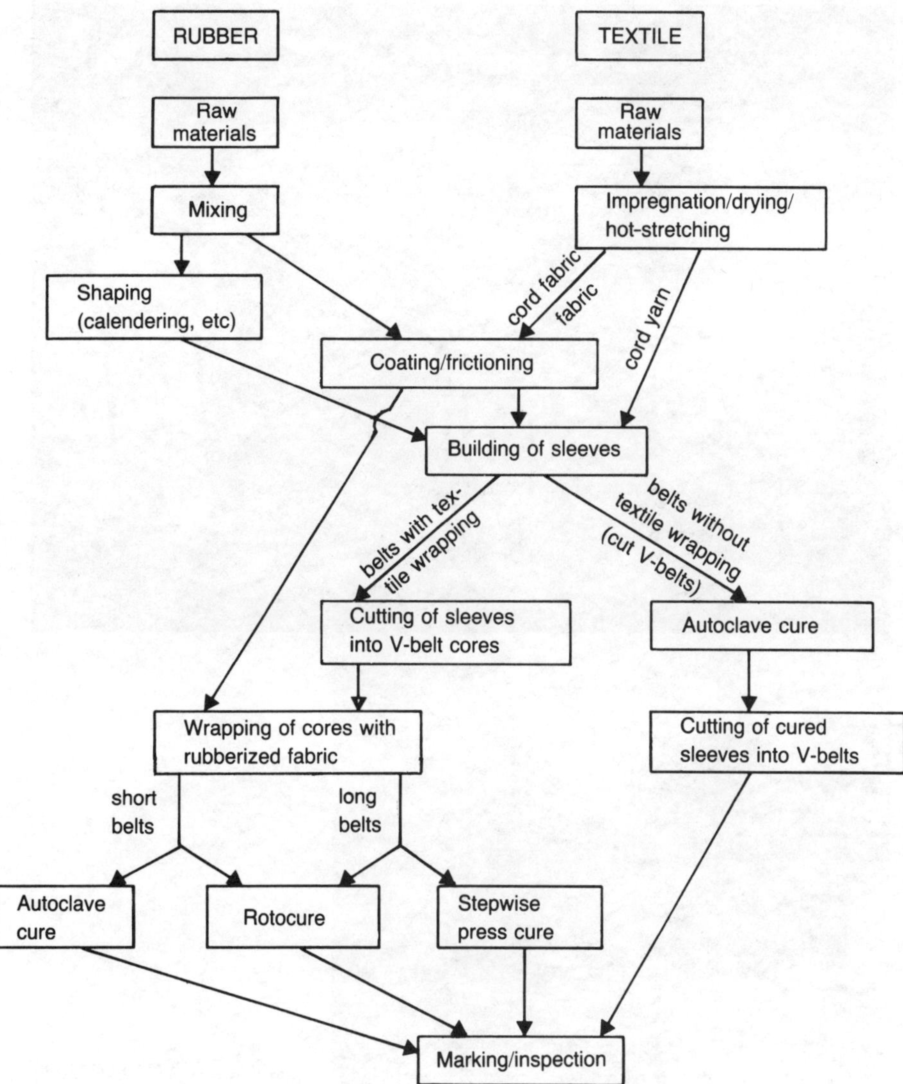

Figure 7-1. Flow sheet for V-belt manufacture

Figure 7-2. Snapshots from the mixing plant of a rubber factory
A. *Working on a roll mill below the internal mixer*
B. *Control room from which the process is supervised from raw materials transportation and weighing to removal of the ready-mixed slabs*

mixed in and well distributed and dispersed in the mix. When dumping the batch it has usually reached a temperature of 120–160°C. When the batch has been cooled down, the ingredients of the curing system are added in a second mixing step. When dumping the final mix its temperature should not exceed 100°C to avoid scorching (incipient vulcanization) of the material. After mixing the batch is given a shape suitable for consecutive operations, such as sheeting out on a rollmill or in an extruder.

Characteristic for rubber processing, as opposed to e g plastics processing, is that materials used are formulated and produced in connection with consecutive operations. By suitable compounding the processing and end properties of the material can be varied to exactly fit the requirements of the end product. A considerable part of the investments in a rubber factory lies in the mixing department. The handling of many additives involves health hazards, meaning that handling in closed systems is increasingly used. The end properties of the material depend largely on good distribution of all the ingredients added. Furthermore, fillers, especially reinforcing fillers, should be well dispersed. The individual, extremely small particles of reinforcing fillers are normally agglomerated and pelletized. To get the desired reinforcing effect of small particles, pellets and agglomerates have to be broken up during mixing into small aggregates or individual particles – a process termed dispersion.

7.3 Textile processing

Material for textile reinforcement is delivered from the textile industry to the rubber industry in the form of cord yarn, cord fabric or other types of fabric. These have to be impregnated with bonding agents that will lead to good adhesive bonding between textile and rubber during vulcanization. The most widely used bonding system for this purpose is a mixture of rubber latex and resin, called RFL (Resorcinol-Formaldehyde-Latex). The choice of proportion and type of these ingredients has a decisive influence on the adhesive bonding to various rubber materials. Sometimes the RFL dip is complemented with other bonding systems, e g based on isocyanates. Textiles already treated with bonding agents can be supplied to those customers, who have no equipment of their own. During drying after impregnation the textile material is usually exposed to hot-stretching, i e stretching at elevated temperatures, figure 7-3, cf section 6.5.4. During hot-stretching water or solvents from the impregnation liquids are driven off. The textile material is stretched and oriented leading to increased tensile strength and tensile stiffness. Proper

Figure 7-3. The properties of the textile reinforcement material are carefully optimized by balancing time, temperature and tension during the hot-stretch procedure after impregnation with materials that will promote adhesion

balancing of time, temperature and stresses or strains during hot-stretching is of great importance for the end properties of the textile reinforcement, cf figure 6-13.

7.4 Shaping

After mixing and textile processing the materials will be shaped and combined in a way suitable for the following building operations. A common shaping method is *calendering*, figure 7-4, which means sheeting out the material by letting it pass through a series of roll-nips. Through calendering sheeting, film or laminates of one or several rubber materials can be produced. The machinery is also used to manufacture fabrics impregnated and coated with rubber and the process is then termed *coating* if the rolls in the roll-nip run at the same speed and *frictioning* if they run at different speeds. By frictioning the rubber is "smeared" into the fabric, giving better penetration and fill-out than with coating.

108

Figure 7-4. In a coating calender textile fabric will be coated by rubber when passing through a series of roll nips giving coated fabric which in V-belts for instance can be used as reinforcement, stiffening or wrapping

Another common shaping method is *extrusion,* whereby long profiles or sections can be produced by forcing the rubber with the aid of a screw through a die. The method can for instance be used to produce bottom rubber sections or to cover cords or cable cords with rubber.

7.5 Building

The next step in the manufacture of V-belts is to join the various components together, starting from the pre-shaped materials to either V-belts or sleeves from which V-belts can be cut. Usually a cylindrical sleeve is first built, figure 7-5. For belts of short lengths the sleeve can be built on one drum, while

Figure 7-5. Building of sleeves for V-belts or V-belt cores on rotating drum

sleeves for long belts have to be built between two drums. Starting from calendered materials top rubber, adhesion rubber, reinforcement and bottom rubber are here joined together. As the sleeve, during the building operation, is exposed to considerable tensile forces, it is usually built inside out, i e with the reinforcement as close as possible to the surface of the building drums. In this way the reinforcement is prevented from cutting down into the underlying thick bottom rubber.

In the manufacture of cut V-belts the sleeves from the building operation pass directly to the vulcanization step. In the manufacture of V-belts with textile wrapping the sleeves are first cut up into individual V-belt cores. These are then wrapped with the rubber-coated textile in a special machine, figure 7-6.

In a not so widely used alternative method of manufacture belts are built individually. Extruded profiles of the bottom rubber are spliced to endless rings which are then wound with reinforcement cord, usually embedded in extruded-on adhesion rubber. The V-belt cores thus obtained are then wrapped with rubber-impregnated textile wrapping. The process can be largely automated.

110

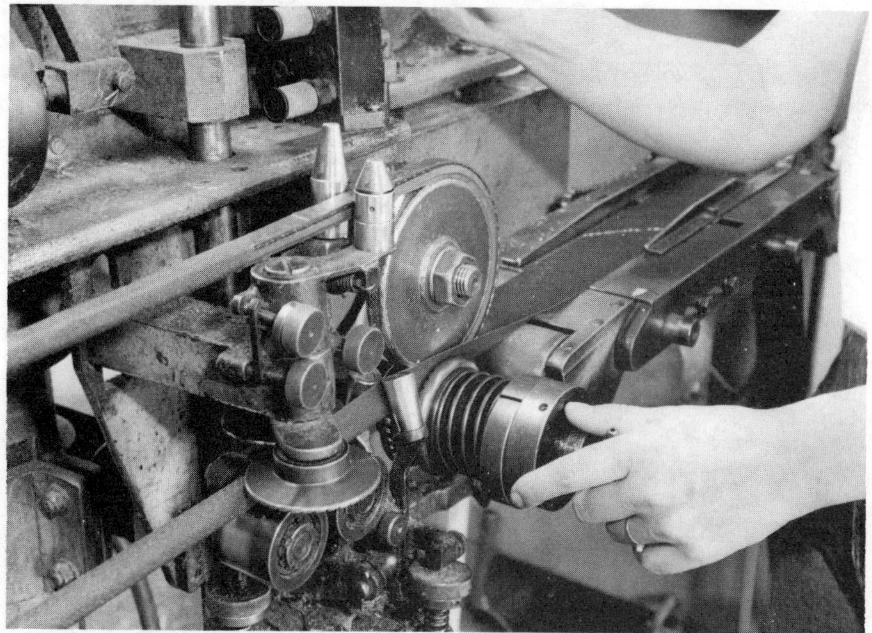

Figure 7-6. Wrapping of unvulcanized V-belt core with rubberized fabric

7.6 Vulcanization

Vulcanization occurs by heating the unvulcanized rubber to 140–180°C. The curing times necessary depend on temperature and the thickness of the section to be vulcanized. Thick sections require longer heating times to bring the material up to vulcanization temperature. It is highly desirable to have the material under pressure during vulcanization, in order to get the various rubber materials to flow together, eliminate porosity and contribute to good adhesion. Sometimes sleeves or individual belts are cooled in the vulcanization pan mould after vulcanization in order to avoid that porosity develops and to keep the textile material stabilized during the critical high temperature period.

For manufacture of cut V-belts the complete sleeve is vulcanized, usually in an autoclave, figure 7-7. Pressure is exerted on the inside of the sleeve with the aid of a steam-heated curing bladder which presses the sleeve against the steam-jacketed autoclave mantle. If a cogged underside on the belt is desired, a profiled, vulcanized rubber sleeve (diaphragm) is placed on the inside of

Figure 7-7. Cured sleeve for the manufacture of cut V-belts is lifted out of the vulcanization autoclave

the sleeve to be cured. When pressure is applied from the curing bladder the wave pattern from the cured sleeve will be pressed into the sleeve to be cured. A curing bladder placed on the outside of the sleeve can also be used which presses the sleeve during curing against a cylinder on the inside. The profiled diaphragm is also in this case placed between the curing bladder and sleeve to impart the desired profile. Belt length is limited by the available diameter of the autoclave and is usually below 5 m. This limitation is not so serious as cut V-belts with cogged undersides are best suited for compact drives involving high bending frequency and small pulleys. After curing the sleeve is cooled in the autoclave before being removed. Cooling time may be as long as the curing time.

Short wrapped V-belts are also usually cured in a steam autoclave, figure 7-8. The ready-built V-belts are then first mounted in a mould composed of dismountable rings, which form the desired profile when mounted together. During curing, pressure is applied on the outside of the mounted V-belts from a steam-heated curing bladder.

Long wrapped V-belts are cured step-wise in a moulded open-sided press, figure 7-9. During curing the V-belts are kept under tension from tension

112

Figure 7-8. Vulcanization of short length V-belts in vulcanization autoclave
A. The belts mounted on the mould are lowered into the autoclave
B. After vulcanization the mould is dismounted and the cured V-belts removed

Figure 7-9. Step-wise vulcanization of long length V-belts in hydraulic press under tension from tension rollers

rollers and moved step by step towards the press platens where curing is performed by application of high pressure and usually in steam-heated moulds.

A curing method of growing importance, which can be used for short as well as for long belts is the rotocure. The equipment involves a slowly rotating, internally heated steel drum, grooved on the outside to correspond to the desired belt profile. The belts are pressed down in the grooves by a steel or textile flat belt and are continuously fed through the curing zone (the contact zone with the drum) by feed rollers and the rotation of the drum, until the whole belt is cured.

7.7 Cutting up of V-belts

Cut V-belts are cut off from the cured sleeves, figure 7-10. Close tolerances and accurate belt angles must be maintained.

114

Figure 7-10. Automatic cutter for manufacture of cut V-belts from cured sleeves

7.8 After-treatment

After curing the following operations may be necessary:

Trimming, e g deflashing.

Visual inspection to reveal any macroscopic defects.

Dimensional checking, mostly only a measurment of belt length, figure 7-11, but this may also include checking of the profile section dimensions and the belt angle, cf section 8.2.

Marking for easier identification, using the principles discussed in section 10.4. Technically marking can be performed with the aid of marking inks, labels or marking tape.

Figure 7-11. Visual inspection and measurement of belt length of V-belts after manufacture

8. Testing

8.1 Introduction

In this chapter the various tests will be described that are used to evaluate the properties of V-belts. For the control of delivered belts dimensional control and visual inspection are usually satisfactory. For evaluation of new types of V-belts, or new applications for V-belts, tests for determining power rating or fatigue properties may be necessary, performed either in special testing machines or in service installations under normal operating conditions. In special cases other tests may be necessary, e g to determine electrical resistivity or flammability.

8.2 Dimensional control

The most important dimensional control is checking of belt length. This control is especially of importance for belts to be used in multiple belt drives. Too large differences in belt length will here lead to uneven load distribution and shortened belt life. Different ways of defining and measuring belt lengths are discussed in section 11.2.

Measurement of belt lengths is usually made by placing the belt over two standardized pulleys under a prescribed tension, figure 8-1. Belt length can then be read off on a scale.

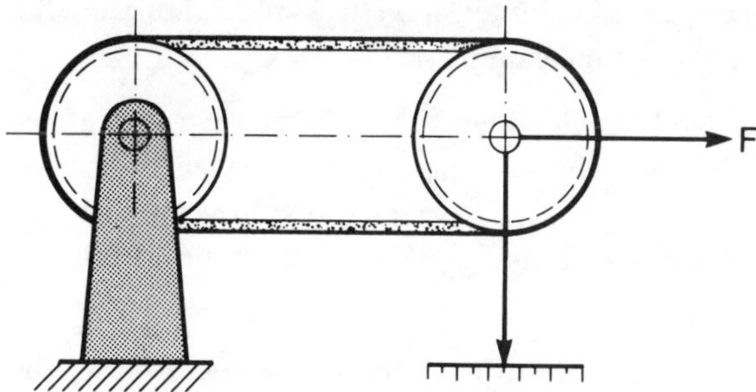

Figure 8-1. Inspection of pitch length or datum length of V-belt. The belt is placed over two identical, standardized pulleys and loaded with measurement tension force F, after which length can be read off from a graduated scale

Pulley grooves are standardized and their dimensions should be checked carefully, cf section 5.2.1. Correct belt profile dimensions can be studied by observing how deep the belt sinks into the pulley groove. Details of methods used for this measurement are presented in different standards, cf Appendix C.

8.3 Tests for the determination of power rating and fatigue life

8.3.1 General

For testing the durability (working life) of V-belts tests simulating actual operating conditions are to be preferred. V-belts are then mounted in a drive containing two or more pulleys, one of which is driven while the others are either loaded (braked) or running idle. The most important factors to vary are belt tension, belt speed and diameter of pulleys, cf chapter 15. In testing machines belt speeds lower than optimal (i e lower than the speed that gives the highest power rating) are usually preferred to avoid too heavy, too power-consuming and too expensive pieces of equipment. The severeness of the test is preferably increased by using higher belt tension and small-diameter pulleys and transmitting higher power ratings than normally recommended for the chosen belt speed and pulley diameter. In belt testing machines only one belt is usually tested at a time on each pulley, even though multiple belt drives may be more typical in practice.

Using testing machines with loaded (braked) driven pulleys power rating can be determined in short duration tests by for instance measuring power transmitted and slip as a function of belt tension. In long duration tests the fatigue properties of the belts can be studied. In testing machines with fixed centre distance the gradual decrease in belt tension (stress relaxation) and the gradual increase in slip can also be measured.

In fatigue tests belts are loaded heavier than normally, to give belt life of usually 100–500 h, so that the tests can be completed in a reasonable period of time. This means a shortening of normal, expected belt life by a factor of about 100, which, however, is considered acceptable to maintain correlation with performance in service.

In fatigue testing of V-belts it is important to check the temperature of the surrounding atmosphere. An increase of the surrounding temperature by 20°C usually corresponds to an increase of belt operating temperature of 10°C,

118

reducing belt life by approximately one half, cf section 6.4.1. Good ventilation and/or cooling of the air may be necessary to avoid the temperature rise that may be expected as a result of the energy spent during the testing.

8.3.2 Testing machine with no power transmission

These machines have one driving pulley and one driven pulley running idle with no power transmission, but loaded with weights to give tension in the belt, a so-called dead-weight machine. The operating principle is indicated in figure 8-2 and a photo of a typical machine is found in figure 8-3. As there is no braking, measurements of slip and power rating cannot be performed. Fatigue of the belts due to their bending over the pulleys can be studied and also their creep under influence of tension forces and dynamic flexing.

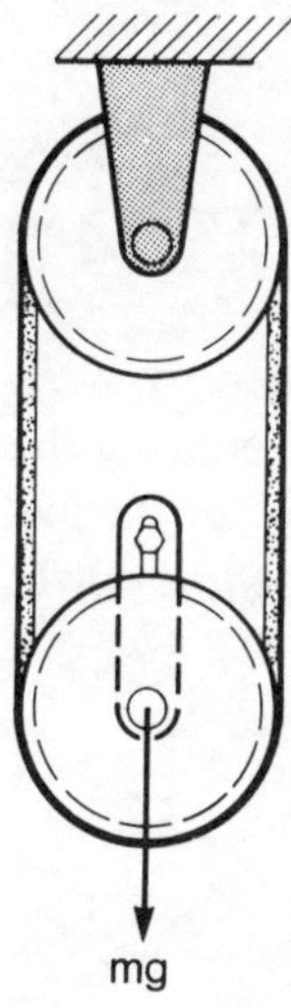

Figure 8-2. Principal design of fatigue testing machine for V-belts with dead-weight load

Figure 8-3. Machines for fatigue testing of V-belts with a fixed load but with no power uptake at the driven pulley (dead-weight machines)

119

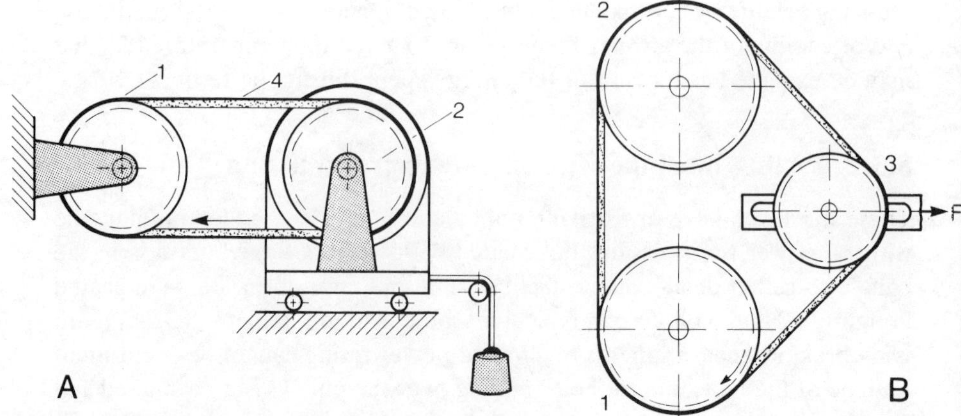

Figure 8-4. Principal design of fatigue testing machines for V-belts with power uptake at the driven pulley
1. driving pulley, 2. driven, braked pulley, 3. idler pulley, 4. V-belt

A. Without idler pulley
B. With idler pulley for testing of automotive belts according to ISO 5287

Conditions are thus different from the normal operating case, reducing the correlation between test data and practice. The method is justified by the low cost of the equipment used and its low operating cost due to the low power consumption involved.

8.3.3 Testing machine with power transmission

To better simulate the practical operating case, testing machines that transmit power are used, i e with one driving and one driven (braked) pulley, figure 8-4. This will also make measurement of slip and power rating possible. As a result of the power transmission a pulsating tensile load between the tight and slack side of the belt is obtained, which corresponds to a practical case and leads to better correlation of fatigue life with service installations.

One type of equipment has driven pulleys that are electrically braked, figure 8-5. The driving motor is usually fed with direct current and the electrical brake constructed as a generator, which feeds its current back to the driving motor, thereby reducing the power consumption of the equipment. Machines of this type are built for all the power ratings that can be used for single V-belts.

120

Figure 8-5. Machine for fatigue testing of V-belts with electrically braked, driven pulley

Figure 8-6. Machine for fatigue testing of V-belts, primarily automotive V-belts, with hydraulically braked (water pump), driven pulley

Another type of equipment has a hydraulic brake on the driven pulley, figure 8-6. It is mainly used for evaluating and testing of automotive belts. Idler pulleys are commonly used to control belt tension and for measurement of creep during the test. The hydraulic brake is usually a water pump where the transmitted power is used to heat the water and thus lost. Detailed descriptions of the test procedure are to be found in the standards, e g ISO 5287.

8.3.4 Field tests

The fatigue testing of V-belts in laboratory testing machines is made under such conditions that the working life of the belt is roughly only 1% of the working life expected in service. Many factors in service, as for instance surrounding atmosphere or load variations, are difficult to simulate in a laboratory test. The most reliable test results are therefore to be expected from service tests. Their main drawbacks are the long testing times involved and the practical difficulty of following and observing the changes appearing during the test. Service tests have to be carefully planned and their results closely observed. Statistics on customer complaints may also be used to get an idea of working life of a certain type of V-belts in various applications.

8.4 Other tests

Examples of other tests, occasionally used for V-belts, are among others the following:

Testing of rubber materials: The correct composition of rubber materials can be checked by chemical analysis. Test specimens can also be cut from V-belts for instance for determination of hardness, tensile strength, elongation at break, ozone resistance, cold resistance, ageing resistance or swelling in liquids, cf section 6.4.1. These tests are mainly applied to check that correct materials of correct composition are used, e g when complaints are obtained, or to keep an eye on competitors' products.

Testing of textile materials: This may include cutting up of the belt for identification of the textile reinforcement used. It is important to check the exact placement of the reinforcing cords, as incorrect placement may lead to uneven stress distribution and drastically reduced working life. Test specimens may be cut from the belts to measure adhesion between textile and rubber. Individual cords can be removed for testing of stiffness (modulus), strength and elongation at break. Static stress-strain tests can be performed on a length section of a belt, but test results should be used with care as there is usually very poor correlation between static tensile strength and the dynamic strength (fatigue resistance) of a belt. The type of textile fibre can be determined by chemical analysis. These test methods are mainly used to check that the V-belts are correctly constructed, e g when complaints are obtained and to keep an eye on competitors' products, cf section 6.5.

Electrical resistivity: Slipping and rubbing of V-belts against pulleys may lead to a build-up of electro-static charge. This can lead to spark discharges, which is undersirable for instance when explosive atmospheres, e g containing solvent vapours, are at hand. It is then required that "antistatic" rubber materials are used, i e having a certain electric conductivity, large enough for electro-static charge to be dissipated. This is usually tested on V-belts mounted between pulleys as in ISO 1813.

Flammability: In many applications it is important that a fire, starting at one point of the drive, will not spread along the belt. The flammability is usually tested on complete sections of the belt as in British standard BS 3790.

9. Installation and maintenance

9.1 Introduction

In this chapter some of the practical problems encountered when using V-belts and V-belt drives will be discussed. This includes installation of pulleys and V-belts, checking and upkeep of proper belt tension, joining procedures, safety devices, problems in service and how they can be overcome, and finally recommendations for the storage of V-belts.

9.2 Installation

The basic requirement for good belt drive operation is that pulleys are firmly mounted on the shafts and well aligned in the same plane. Pulleys with integral hubs are mounted with keys and keyways like other wheels, figure 9-1. A special pulley drag is required for dismounting. Pulleys with dismountable bushings (bushing pulleys) are mounted on the shaft with the aid of keys and keyways, or by clamping with screws, figure 9-2. Screws can also be used for dismounting, so that special pulley drags are not needed. For smooth and stable running as well as for long belt life it is required that pulleys run in the same plane. Axial or angular displacements exceeding $\pm 1°$, e g depending on disalignment of shafts, should not be accepted. The simplest way of checking is with the aid of a steel rule or cord, held tight between the pulleys and close to their sides. Visual inspection of variations in slot width between steel rule/cord and pulley sides will reveal the existence of radial or angular displacement, figure 9-3. Even initially correctly aligned pulleys may give problems in service due to manufacturing defects, poor attachment or wear, figure 9-4. The problems may be experienced as eccentricity of pulley grooves in relation to shaft, fluttering of the pulley due to unstable attachment or wear between hub and shaft or between bushing and shaft or pulley. In addition problems may arise due to axial movement of pulley on the shaft due to unstable attachment or wear between hub or bushing and shaft.

Figure 9-1. Mounting of pulley with integral hub on the shaft. The key is first placed in the keyway after which the pulley is set in place. For small pulleys a copper hammer can assist this operation while larger pulleys may have to be warmed up before mounting on the shaft

Figure 9-2. Mounting of pulley with removable hub (bushing pulley) on the shaft

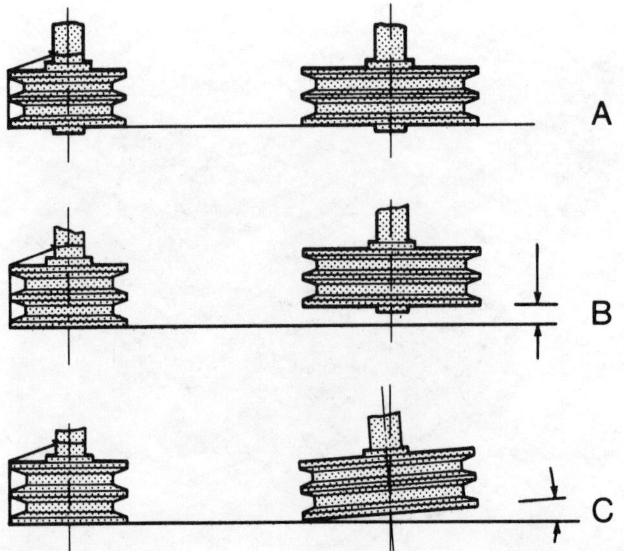

Figure 9-3. Inspection of alignment after mounting. A cord or ruler is kept close to the plane surfaces of the pulleys and the existence of gaps denoted
A. Pulleys well aligned
B. Pulleys axially displaced
C. Pulleys with angular displacement (e g shafts not parallel)

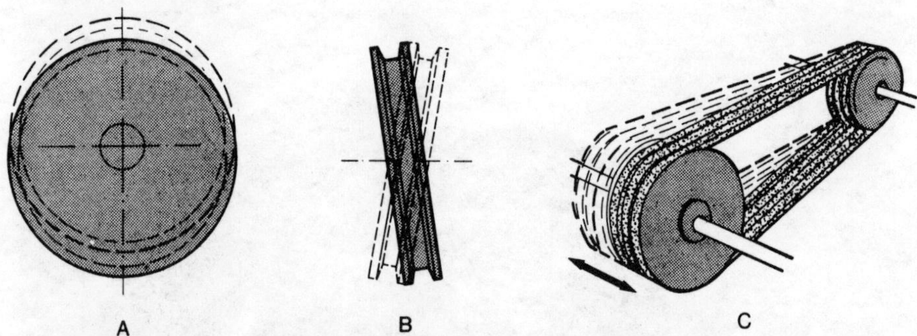

Figure 9-4. Eccentricity (A), wobble (B) and axial movement (C) of V-belt pulleys

The next step in installation is the placement of the V-belt in the pulley groove. The main rule is to avoid prying of the V-belt into the grooves with the aid of tools that can cause damages to the pulley groove as well as to the V-belt. The V-belt should further not be rolled on with the driving shaft in rotation, which may cause injuries to the operator by trapping his hands or fingers between belt and pulley. The correct installation procedure is to make use of the devices for shaft centre line distance adjustment, which are always at

126

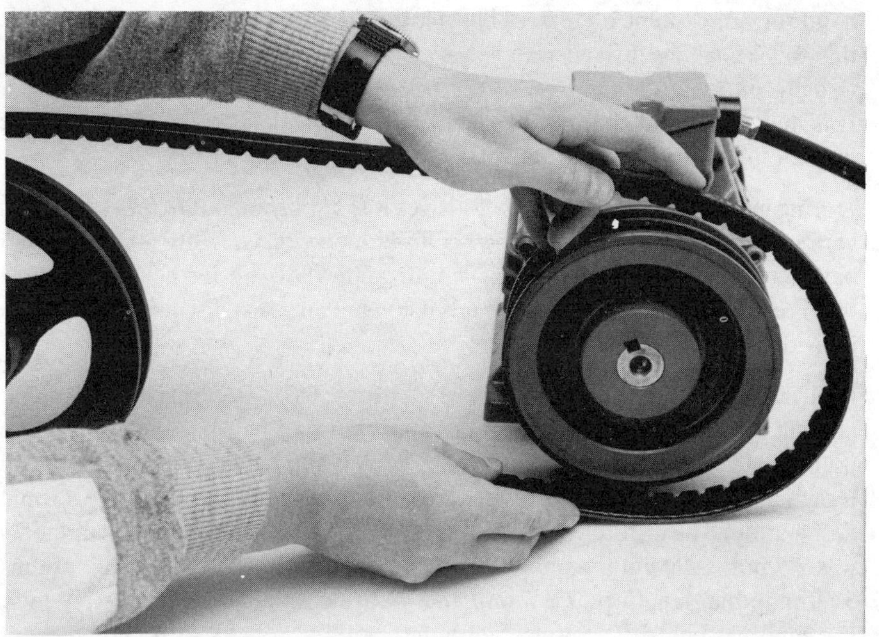

A

B

Figure 9-5. Recommended (A) and not recommended (B) procedure for mounting of V-belts on belt drive. Shaft centre distance should be reduced by utilizing the installation allowances so that the belt can easily be placed in the groove without prying or application of large deformations. Tools used for prying can damage belts as well as pulleys

127

hand for adjustment of correct belt tension, figure 9-5. After installation in this way checking that correct belt section has been used is easily made by visually studying the position of the belt in the pulley groove, cf figure 5-2. It is thus important that the belt neither protrudes above the circumference of the pulley nor that it bottoms in the groove.

During installation of multiple belt drives it is important not to mix belts from various manufacturers or old, used belts with new ones, figure 9-6. Old, used belts have usually increased their length somewhat and further wear makes them run deeper in the grooves, leading to uneven stress distribution between individual belts in the drive.

Earlier manufacturing technique gave such large tolerances on belt length that all manufactured belts of a given nominal standard belt length could not be used together in a multiple belt drive. The belt length was then characterized by a matching number that was marked on each belt in addition to the normal designation. The difference in length between two adjacent matching numbers was 2,5 mm. Matching number 50 corresponded to nominal length and thus 51 to nominal length plus 2,5 mm and 49 to nominal length minus 2,5 mm. For multiple belt drives up to belt lengths of 2,5 m belts with two adjacent matching numbers were permitted and for belt lengths exceeding 2,5 m three adjacent matching numbers. With improved manufacturing technique belt length tolerances are nowadays so small that the need for matching numbers has disappeared and thus all belts of a given standard length can be used in matched sets.

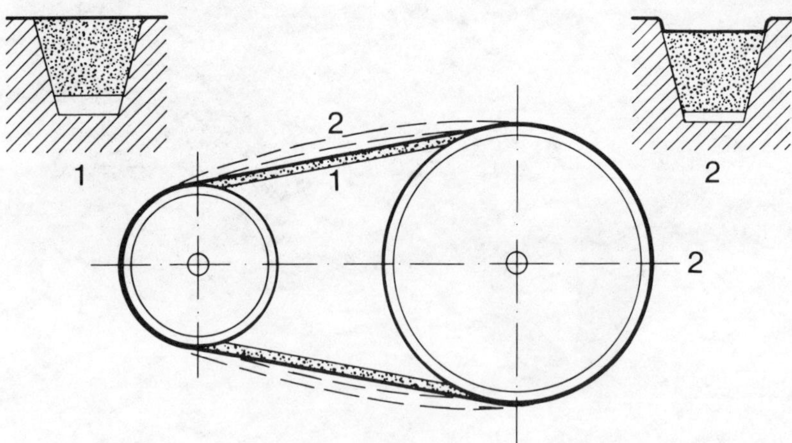

Figure 9-6. Problems are usually experienced in mounting of new (1) and used (2) V-belts together in a multiple belt drive. The used belt can be so worn that it bottoms in the groove

9.3 Belt tension

9.3.1 Choice of correct belt tension

Belt tension should neither be too high nor too low. Too low belt tension leads to excessive slip, reduces power transmission capability and increases wear. Too high belt tension will shorten the life of the belts and increase the loads on shafts and bearings. It can therefore be stated that the correct belt tension is the lowest one that doesn't lead to excessive slip. The belt tension required increases with the size of the V-belt section and varies, with exception for the very largest belt sections, normally within the range 200–2 000 N. This corresponds for belts reinforced with polyester cord to a strain of around 0,5–1,5%, figure 9-7, cf section 14.2.

Profile	Reinforcement	Modulus of elasticity, MPa	Belt section area, mm^2	Recommended pre-tension, N	Calculated belt elongation at recommended pre-tension, %
SPZX	polyester cord 1100×2×3	780	55	200–420	0,5–1,0
SPAX	polyester cord 1100×2×5	690	90	300–650	0,5–1,0
SPBX	polyester cord 1100×3×5	630	150	600–1050	0,6–1,1

Figure 9-7. Example of tension stiffness (modulus of elasticity) for some V-belt profiles together with recommended pre-tension and corresponding elongation obtained (values from Trelleborg AB)

9.3.2 Belt tensioning devices

Various types of belt tensioning devices and their principles of operation are discussed in chapter 14. A short summary is presented in figure 9-8. The devices can either operate with a fixed, but adjustable shaft centre distance or with a constant tension load acting under the whole operating life of the belt.

Method	See section	Readjustment required	Rotation speed can be reversed	Principle
Constant elongation	14.3	yes	yes	The belt is stretched and retained at a constant elongation
Constant load	14.4	no	yes	A constant force is exerted on the belt, e g by springs or weights
Idler pulley	14.5	sometimes	no	The belt is stretched to a constant elongation or loaded with a constant slack side tension by idler pulleys
Weight of driving unit	14.6	no	yes	The weight of the driving unit keeps the belt under constant tension
Self-adjusting motor mounting	14.7	no	no	Belt tension is automatically controlled in proportion to the torque loading
Axial force from split pulley	14.8	no	yes	Belt tension kept at a fixed value or varied in proportion to the torque loading by controlling axial force in a split pulley

Figure 9-8. Different methods for pre-tensioning of V-belts

9.3.3 Adjustment and measurement of belt tension

For adjustment of belt tension the correct value is usually checked with the aid of a so-called tensiometer, figure 9-9. In this simple-to-operate instrument the force needed to deflect the belt in the middle of its span, usually 1/64 of the span length, is determined. The measurement is simple if the instructions accompanying the instrument or given in V-belt catalogues are followed. Examples of some recommended tensiometer values for various V-belt sections and service conditions are presented in figure 9-10. For conversion of tensiometer values to belt tension forces, see section 14.2.

130

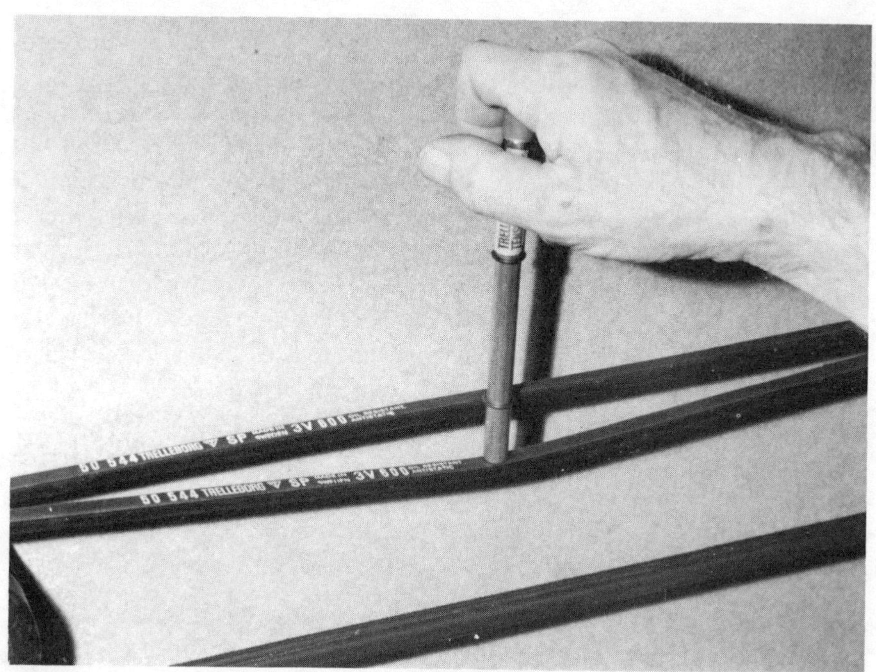

Figure 9-9. *Measurement of belt pre-tension with a tensiometer*

Profile	Small pulley diameter, mm	Recommended deflection force per belt, N	
		Min	Max
SPZX/ 3 VX	56	13	19
	63	15	21
	67–71	16	23
	75–80	17	25
	85–95	19	27
	100–125	22	32
	132–180	26	39
AX	71	16	23
	75–80	17	25
	85–90	19	28
	95–106	20	29
	112–180	22	32

Figure 9-10. Examples of recommended tensiometer values for some V-belt profiles to obtain proper pre-tension. The recommendations are valid for a speed ratio of 2-4:1 and for a rotation speed of the small pulley of 1200–3600 rpm (data for TrellPower X V-belts in catalogue from Trelleborg AB)

9.3.4 Readjustment of belt tension

A V-belt, installed with correct belt tension, will creep or relax in service. If it utilizes fixed shaft centre distance this will lead to a decrease (relaxation) of belt tension. The effect is most marked during the first hours of service but then tends to level out, cf figure 6-14. The belt starts to slip if belt tension falls below a certain value. This requires re-tensioning of the belts after a running-in period of 4–24 h. Further belt tension must be checked at regular intervals, for instance using a tensiometer, and belt drives regularly inspected visually to detect any excessive slip. Wear of pulley grooves or belts will also lead to reduction in belt tension when tensioning to fixed shaft centre distance. In belt tensioning devices working with constant belt tension the need for readjustment disappears.

9.4 Joining

Joining of V-belts using mechanical fasteners, figure 9-11, is seldom used, as the fasteners introduce a weak point in the construction, leading to decreased power ratings or shortened life. As the reinforcement in the belt is usually placed in its longitudinal direction it offers limited anchorage for the mechanical fasteners, which easily eat through the rubber under the influence of tensile and bending forces. Fabric reinforcement with plenty of reinforcing threads also in the weft direction, i e transversal to the length direction of the belt, offers improved anchoring for mechanical fasteners. Mechanical joining of V-belts is only used where installation is facilitated and where a short life or low power ratings are accepted. V-belts with regularly placed pre-drilled holes are then often used to facilitate application of the fasteners, cf figure 3-25 E. For flat belts joining with mechanical fasteners is still common and usually gives good results due to the large width available for joining as compared to V-belts.

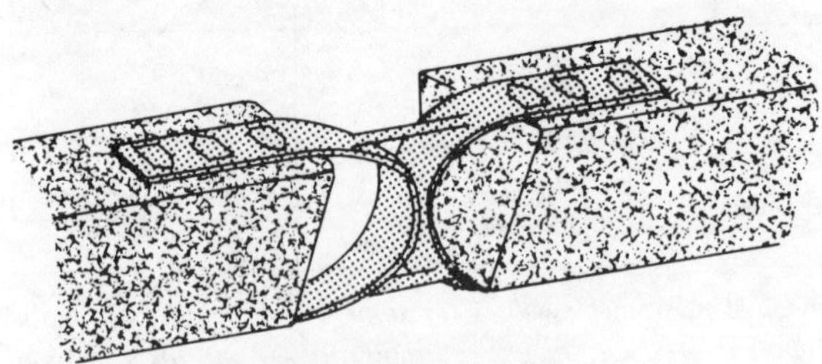

Figure 9-11. Mechanical joining of V-belts

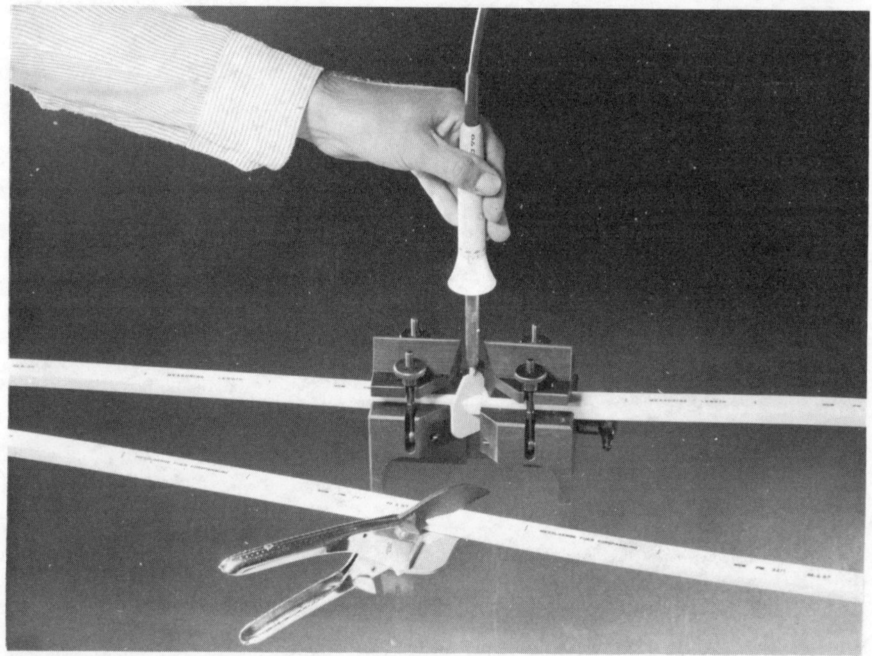

Figure 9-12. Heat-welding fixture for joining of thermoplastic V-belts. A clean cut is made in the belt and the two surfaces to be joined are mounted opposite to each other, melted by electric heating and then rapidly pressed together

Hot welding is another method of joining, that can be used for V-belts manufactured from thermoplastic materials, e g polyether-polyester thermoplastic elastomer (Hytrel). The surfaces to be joined are then heated and rapidly pressed together after which the material solidifies and forms a homogeneous joint, figure 9-12. The unreinforced V-belts that can be joined in this way usually have low power ratings and wear rapidly, which limits the application of this technique.

9.5 Safety devices

Belt drives, like all equipment with moving or rotating parts, should be equipped with safety devices that prevent operators from getting injured, e g by getting hands or fingers trapped between belts and pulleys. Various types of guards, including nets, gratings or casings, may be used, figure 9-13. It is

Figure 9-13. Protective housing at belt drive to act as a guard and prevent human contact with moving parts

important that the guards not be so tight as to prevent ventilation as increased operating temperature of the belts will decrease their life considerably. In dirty or dusty environments tight enclosures are often required to protect the belt drive. These must, however, be constructed to permit good ventilation to be maintained. During inspection, readjustment of belt tension, etc, when guards have to be removed, the belt drive should be shut off. Under all circumstances direct contact between hands and moving belts should be avoided.

9.6 Measures against operating disturbances

A summary of typical disturbances that can be experienced in a V-belt drive is to be found in figure 9-14, giving possible reasons for the disturbances and suggesting suitable measures to overcome the problems. The scheme cannot be totally complete but may be of help in the most typical cases. Most types of disturbances cited require no further discussion or illustration. Typical rupture failures of V-belts are illustrated in figure 9-15. Damages to and wear

Trouble area	Causes	Corrective action
1. Rupture or separation	Overload	Redesign belt drive, change to belts with higher power rating
	Too high belt tension	Reduce belt tension
	Too small pulleys	Change to larger pulleys or to belts with cogged underside
	External mechanical action	Improve shielding (guards)
	Too high operating temperature	Reduce operating temperature (cf 8)
	Injured, worn or clogged pulleys	Change pulleys, improve shielding
	Uneven load distribution from belts of varying age or make	Change to new, matched belts
	Cord yarns ruptured during installation	Change to new belts and install them properly
	Uneven load distribution due to misalignment of shafts	Align the shafts
	Dead stop of driven unit	Find cause and correct
2. Abnormal wear (belt sides)	Wrong, injured or worn pulleys	Change pulleys and, if necessary, belts
	Too small pulleys	Change to larger pulleys or to belts with cogged underside
	Overload	Redesign belt drive
	Belts of varying age or make are used together	Change to matched belts
	Excessive belt slippage	Reduce slippage (cf 10)
	Very old or improperly stored belts	Change to new belts

Trouble area	Causes	Corrective action
	Too high operating temperature	Reduce operating temperature (cf 8)
	Misalignment of drive	Correct alignment
3. Abnormal wear (topside of belt)	Tilting of the belt	Reduce risk of tilting (cf 7)
	Mechanical contact with surrounding objects, e g guards	Remove obstructions
4. Transversal cracking (usually from the bottom)	Too small pulleys	Change to larger pulleys or to belts with cogged underside
	Improper idler pulley	Use idler pulley of sufficient size properly installed
	Material hardened depending on ageing	Reduce operating temperature (cf 8)
	Material occasionally hardened due to the influence of extremely low external temperature	Heat up the belt before starting up
	Foreign material embedded in belt forms starting points for cracks	Improve shielding
5. Tacky or soft surface (often combined with separations or open seams in textile wrapping)	Contact with oil or grease	Clean belts, eliminate oil spill, change to oil-resistant belts
6. Hard and polished surface (also with shallow cracks)	Overheating due to excessive slippage	Increase belt tension and, if necessary, change to new belts
	Too high operating temperature	Reduce operating temperature (cf 8)
7. Tilting of the belt	Excessive variations in torque loading	Change to joined V-belts
	Cords ruptured during installation	Change to new belts properly installed
	Overload	Redesign belt drive

Trouble area	Causes	Corrective action
	Dirt in pulley grooves	Clean grooves and improve shielding
	Misalignment of drive	Correct alignment
	Improper idler pulley	Install idler pulley on slack side of drive close to driver pulley
	Wrong or worn belts or pulley grooves	Change belts and/or pulleys
8. Overheating of belt	Too high surrounding temperature	Remove heat source, increase ventilation
	Too high belt speed	Use smaller diameter pulleys, increase shaft centre distance, use belts of smaller section
	Excessive slippage of belt	Reduce risk of slippage (cf 10)
9. Heavy vibrations and oscillations of belt	Too low belt tension	Increase belt tension and, if necessary, use idler pulleys
	Large variations in shock loading	Change to joined V-belts
	Misalignment of drive	Correct alignment
10. Excessive slippage	Too low belt tension	Increase belt tension
	Too small arc of contact angle	Increase shaft centre distance
	Overload	Redesign belt drive
	Worn belts or pulleys	Change to new belts and/or pulleys
	Oily belts or pulley grooves	Clean up the drive, reduce oil spill
	Insufficient take-up allowances on shaft centre distance	Change to shorter standard belt length or introduce idler pulleys
11. Noise	Excessive belt slippage	Increase belt tension
	Overload	Redesign belt drive
	Belts bottom in the pulley groove	Change to other belts and/or pulleys
	Misalignment of drive	Correct alignment
12. Incorrect driven speed	Excessive slippage of belt	Increase belt tension
	Incorrect speed ratio between pulleys used	Change pulleys

Trouble area	Causes	Corrective action
13. Overheating of bearings	Insufficient lubrication	Improve maintenance
	Worn or injured bearings	Change to new bearings
	Too high belt tension	Reduce belt tension
	Pulleys too far out on shaft	Place pulleys as close to bearings as possible
	Excessive slippage of belt	Increase belt tension
	Too small pulleys	Change to larger pulleys

Figure 9-14. Examples of common defects and disturbances in belt drives, the probable reasons for them and suggested action for remedy

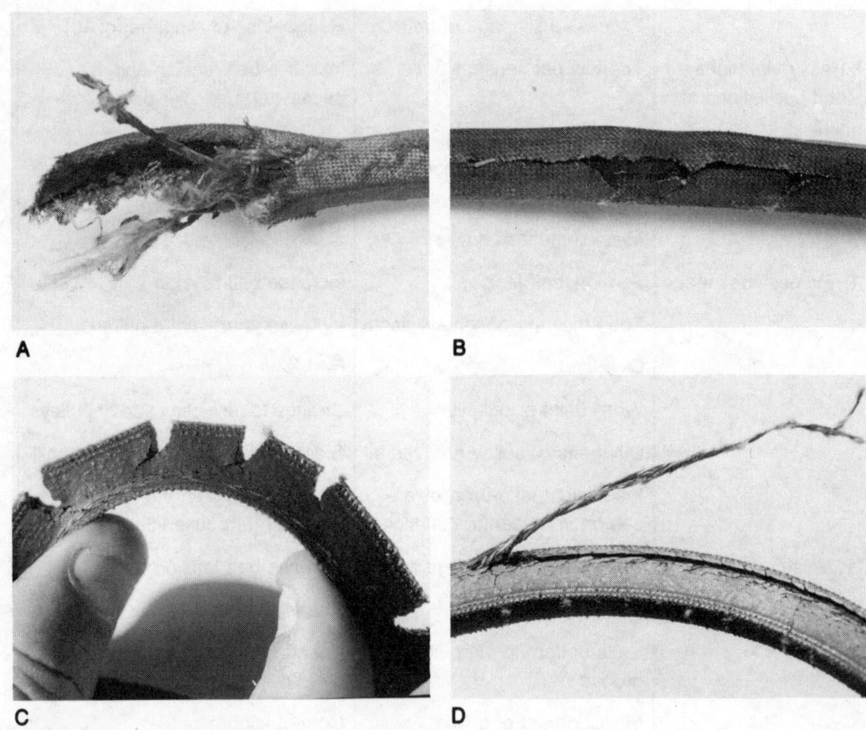

Figure 9-15. Typical fractures in V-belts after inappropriate service
A. Total fracture
B. Delamination
C. Transversal cracks at the bottom
D. Fraying of the reinforcement at the side of cut V-belt

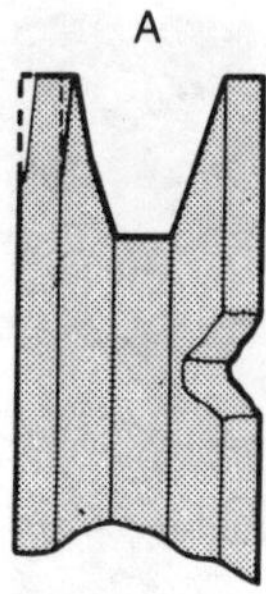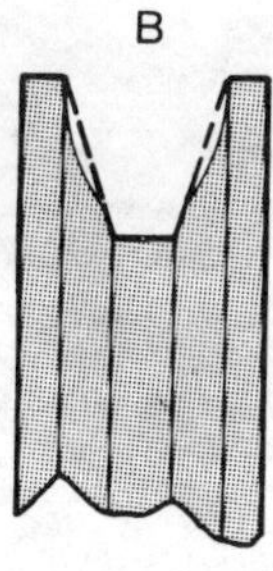

Figure 9-16. Mechanical damages (A) and wear (B) in V-belt pulley grooves

in pulley grooves are demonstrated in figure 9-16. High working temperatures of the belt, e g 80–100°C, can be verified with the aid of instruments. An old rule states that if the belt can be kept with bare hands under at least 5 s after stopping the drive belt temperature is not too high.

The maintenance of flat leather belts included the application of certain resinous materials, so-called belt dressing, that kept the belts flexible and increased their friction against pulleys. For V-belts the use of belt dressing is generally discouraged as it can damage the belts and pick up solid particles that will increase wear of belts and pulleys. In a very few cases belt dressing can give positive effects, e g by increasing the friction of belts with hard and smooth surface and reduce noise during heavy slippage.

9.7 Recommendations for storage of V-belts

V-belts should preferably be stored in dark, cool and dry locations. Exposure to direct sunlight or elevated temperatures will cause some deterioration of the material. Humidity, solvent vapours, acid vapours and high ozone concentrations may also have a deleterious influence, but these are rarely found in such high concentrations as to give problems. During long-time storage V-belts should not be stored in distorted conditions and should preferably be free-hanging over cylindrical rods with a diameter at least corresponding to 10 times the height of the belt section, figure 9-17. With the observation of these rules V-belts can be stored for at least seven years with no risk of deterioration.

Figure 9-17. Recommended method for hanging up of V-belts during storage

140

10. Standardization and marking

10.1 Introduction

Standardization of transmission belts and pulleys is important to facilitate construction, installation, replacement and spare part service. In this chapter the organizations of greatest importance for the standardization of V-belts and pulleys are described, after which the scope of the standards will be discussed in general. Finally the designation and marking of V-belts, which are often related to standardization, are reviewed. Lists if important, relevant standards and a tabulated summary of their main contents are found in Appendix C. As standardization is a continuous process, leading to the issue of new standards each year, data in Appendix C has to be checked when up-to-date information is desired.

10.2 Organizations for standardization

Standardization of V-belts and pulleys, like for many other equipment components, can be pursued at different levels, for instance:

1. International standardization
2. National standardization
3. Standardization within a country, for instance within a branch organization or private company

International standardization has the advantage that agreement on worldwide standards can be reached from the start, which may then be used all over the world without restrictions. The work within the international standardization is, however, often concentrated on conciliation of the ideas and suggestions already developed in national standards. Today most industrialized countries have national standardization organizations which cooperate within the international organization of standardization. In addition there exists a great number of standards from private companies, in Sweden for instance from Volvo, which are formulated to fit the exact needs of the company in question.

The tendency today is that international standardization increases in importance and partly replaces national and private standardization. International standards, when available, are nowadays mostly accepted by national standards organizations to full extent or in a slightly modified form and translated into the national language. This trend favours international trade which has become of growing importance.

For the standardization of V-belts and pulleys the following organizations of standardization are the most important ones:

International Organization for Standardization (ISO), with central secretariat in Geneva, handles international standardization. The work within ISO is carried out in Technical Committees (TC), in which national standard organizations are represented according to their wish. TC 41 "Pulleys and belts (including veebelts)" handles the standardization of transmission belts, pulleys and conveyor belting. The work is divided on Subcommittees (SC), of which SC 1 handles the standardization of transmission belts and pulleys. Work starts with a draft, which after discussion and agreement within the Subcommittee is circulated among the members of TC 41 as a Draft ISO Proposal (DP) for comments and voting of approval or disapproval. When a solid majority for agreement has been reached a Draft ISO Standard (DIS) is issued for circulation and voted upon by all ISO member bodies, after which an ISO Standard can be issued if agreement is reached. The many steps and careful procedure involved requires 5–10 years starting from a Draft before an ISO Standard is agreed upon. The process will, however, lead to a result of good agreement which is technically well qualified. The standard is then designated as "ISO" followed by a 3-digit or 4-digit number. Previously the word "Recommendation" was preferred to "Standard". The Recommendations thus issued are designated with "ISO-R", followed by a 3-digit or 4-digit number. Today about twenty ISO standards are available for transmission belts and pulleys, cf Appendix C.

Standardiseringskommissionen i Sverige (SIS) is the Swedish national organization for standardization. The standardization of transmission belts and pulleys is, as for many other mechanical components, handled by Sveriges Mekanstandardisering (SMS) which collaborates with SIS in this area. The work is divided among technical committees (TK), of which TK 46 handles the standardization of transmission belts and pulleys. About twenty SMS standards are available in this area, cf Appendix C. The standards were earlier developed from scratch, but nowadays work is concentrated on cooperating within ISO and applying the results from the international standardization on the national scale. Older standards are designated by "SMS" followed by a 3-digit or 4-digit number. Modern standards, directly taken from corresponding ISO standards, are designated by "SS/ISO" followed by the ISO standard number.

Deutsches Institut für Normung (DIN) is the German national standards organization, the work of which has been guiding much of the international standardization work on transmission belts and pulleys. The standards are designated by "DIN" followed by a 3- to 5-digit number. Standards with a given number may be divided into several parts ("Teil 1", "Teil 2", etc).

British Standards Institution (BSI) is the British national organization for standardization. Standards are designated by "BS" followed by a 3-digit or 4-digit number. The number of standards for V-belts and pulleys is small, but each standard usually contains several parts.

Rubber Manufacturers Association of America (RMA) is a branch organization which has done a great amount of pioneering work for the standardization of transmission belts and pulleys in the USA, which also forms the basis for much of the international standardization. For standardization of transmission belts and pulleys, RMA today cooperates with two other organizations, viz the Mechanical Power Transmission Association (MPTA) and the Rubber Association of Canada (RAC). The standards are therefore often designated "MPTA-RMA-RAC Engineering Standards". Individual standards are designated IP-20 to IP-26, cf Appendix C. The standards are mainly related to transmission belts and pulleys for industrial drives.

Society of Automotive Engineers (SAE) is the most important organization within the USA, that issues standards for automotive components, including transmission belts and pulleys. Standards are designated by "SAE J" followed by a 3-digit or 4-digit number. SAE has been the leading organization for development of standards for transmission belts and pulleys for automobiles, and their work forms the basis for later national and international standardization.

American Society of Agricultural Engineers (ASAE) develops standards for components in agricultural machinery, including transmission belts. The results of their work have formed the basis for many national and international standards in the area.

10.3 Scope of the standards

The most important scope of standardization is *dimensions,* e g section profiles of belts and pulley grooves, diameters of pulleys and lengths of belts. This type of standardization is a pre-requisite for the fitting of belts and pulleys together and for easy replacement. The standards also give instructions for how dimensions should be measured in an accurate and reproducible manner.

Power rating is given in many standards. Results are usually presented in tables for various, standardized sections with varying rotation speed and diameter of the driving (small) pulley. Power ratings can also be calculated with the aid of standardized equations, cf chapter 16.

Design procedures for belt drives are often found in standards, especially for the simple case with two pulleys operating in the same plane, cf chapter 18. Standards describe the steps in the designed procedure and suggest how corrections should be applied for varying service conditions, arcs of contact angle and belt lengths.

Special requirements may include demands on e g electric resistivity or flammability.

In Appendix C the most important standards for transmission belts and pulleys from the organizations mentioned in section 10.2 are listed, and a table is presented in which the scope of the various standards is indicated.

10.4 Designation and marking

Many standards prescribe how transmission belts should be designated and marked. Most important is to identify profile section, where the designations used are presented in figures 3-2, 3-5, 3-7, 3-9, 3-11, 3-13, 3-18, 3-20 and 3-24. In addition the designation must contain a number giving the standard length of the belt. In international standards this is usually given in millimetres, but older designations based on inches are still widely used. Belts are also often marked with the name of the manufacturing company or the supplier. In addition, various fancy designations may be used as "Premium", "Super" or "HC" (= High Capacity). Examples of markings on some V-belts are given in figure 10-1.

Figure 10-1. Examples of marking of V-belts. The letter M stands for matched V-belts and the following number is a code for date of manufacture
On top: ISO marking with datum length in millimetres and RMA marking with external length in tenths of an inch
Middle: RMA marking with internal length in inches
At bottom: ISO marking alone

11. Geometrical calculations

11.1 Introduction

In this chapter geometrical factors, i e dimensions and calculations of dimensions that characterize belts, pulleys and belt drives will be discussed. First the various ways of characterizing V-belt profiles and lengths, diameters and groove dimensions of pulleys are treated. Then a discussion of geometrical calculations follows, for the simplest type of belt drive with two pulleys in the same plane. Finally more complicated belt drives are discussed, e g these with more than two pulleys or with pulleys in different planes, as well as geometrical calculations of synchronous belt drives.

11.2 Various ways of characterizing V-belt and pulley dimensions

If it is assumed that belts and pulleys are manufactured to exact dimensions without tolerances, that dimensions are not changed by wear, that belts are not elongated under the influence of tension forces and that belt thickness (height) is so small that it can be neglected, then the characterization of dimensions of belts and pulleys would be unambiguous and geometrical calculations simple. For rough estimates it can be assumed that these ideal conditions apply, but for more accurate calculations the factors mentioned must be taken into account and the dimensions defined with greater accuracy.

For characterizing belt profiles, belt lengths, pulley groove profiles and pulley diameters three different systems of measurement are used.

Pitch dimensions are based on the pitch lines, which together form the pitch zone in the belt. The pitch line is a line in a belt which keeps its length constant when the belt is bent. Pitch line is sometimes also termed neutral line, and pitch zone is termed neutral zone. In a section made up of a homogeneous material the pitch zone will be located at the centre of gravity. A V-belt, however, is a composite product with the component of highest tensile modulus – the reinforcement – located close to the top side of the profile. The pitch line and the pitch zone will therefore be located either in the reinforcement or just below it. The width of the V-belt in the pitch zone is termed pitch width w_p. The length of the belt in the pitch zone is termed pitch length L_p. The width of the pulley groove corresponding to the pitch width of the belt is also termed pitch width w_p. The diameter of the pulley up to the pitch width of the groove is termed pitch diameter d_p and corresponding circumference is termed pitch circumference C_p. Pitch dimensions are important for calculation of speed ratio, tension forces in belts and power ratings, but have the drawback that they are difficult to measure directly. They will furthermore depend on tolerances of belts as well as of pulleys and on the exact procedure for their measurements, cf section 8.2.

Datum dimensions are based on the nominal dimensions of the pulleys, firstly the datum width of the pulley groove w_d, the datum diameter d_d and the datum circumference C_d. These dimensions are thus independent of the actual tolerances of the pulleys, which may arise from manufacture or wear in service. Starting from datum dimensions for the pulleys the datum length L_d of the belt is defined. The small difference between datum dimensions and pitch dimensions is usually given as the datum line differential b_d, which is defined as the radial difference between the planes for pitch width and datum width, i e:

$$d_d = d_p \pm 2b_d \quad\text{...}\quad (11{:}1)$$

Datum dimensions were introduced within the international standardization a few years ago and are expected to become increasingly used, even though they are at present relatively unknown. For most practical purposes they can be replaced by pitch dimensions with acceptable accuracy.

Effective dimensions are referred to the outer diameter of the pulley which normally coincides with the top side of the V-belt, and are as a result easy to measure. The largest width or top width w is usually measured when the belts are unloaded. More important is the effective length L_e, of the belt, measured according to the procedure described in section 8.2, i e with a belt under a specified tension force mounted between two standardized pulleys. The diameter in the pulleys, level with the top surface of the belt is then termed effective diameter d_e. The width of the pulley groove corresponding to its effective diameter is termed effective width w_e and the corresponding circumference effective circumference C_e. Most pulley grooves are so dimensioned that their effective width coincides with their top width and the effective diameter of the pulleys with their outer diameter. Deep-groove pulleys are also available where the effective width and effective diameter are below the outer diameter and corresponding top width. The difference between effective dimensions and pitch dimensions is given by the effective line differential b_e which is the radial difference between the planes containing the effective width and pitch width respectively, i e:

$$d_e = d_p + 2b_e \qquad \qquad (11{:}2)$$

An additional way of characterizing dimensions, although not widely used, is to start from the internal length of the belt.

Some of the dimensions described above are illustrated in figure 11-1. For more precise and detailed information on definitions, methods of measurement and application, reference is made to relevant standards. Thus for classical and narrow V-belts RMA uses the effective system while ISO standards are gradually being changed over from the pitch to the datum

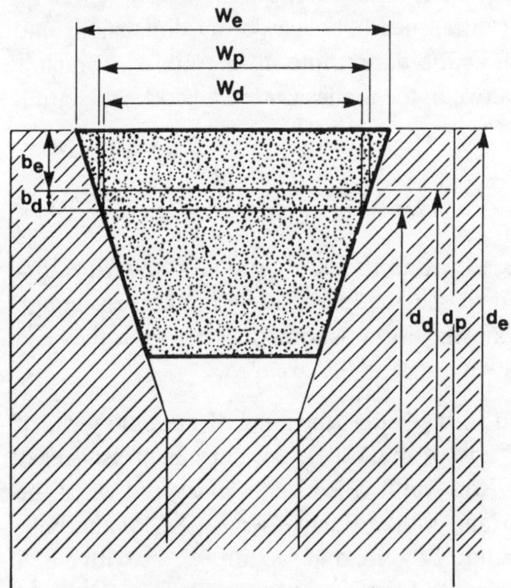

Figure 11-1. Definition of diameter d and width w for V-belt in pulley groove using the effective system (index e), the pitch system (index p) or the datum system (index d). In the figure the radial displacements between the levels where width is measured, viz datum line differential b_d between pitch width and datum width and effective line differential b_e between pitch width and effective width are also given

system. For agricultural variator belts ISO 3410 uses the pitch system while ASAE S211.3 uses for roughly the same belts the effective system. For narrow automotive belts both ISO 2790 and SAE J636c use the effective system of measurement. The delay in introduction of new standards explains why the marking of the belts still can either be in inches or in SI units and can sometimes also refer to the internal length of the belt.

The height of the belt section is usually given as T. The relative height of the belt section is the ratio between belt section height and pitch width, i e:

$$\text{Relative height} = \frac{T}{w_p} \qquad \qquad (11:3)$$

Accurate determination of V-belt dimensions is made difficult by the fact that the belts are constructed from soft and elastic materials and tend to deform under the influence of the forces applied during measurement. It is therefore important that the exact procedure of measurement is well standardized. Best accuracy is obtained by referring the dimensions to the pulleys. In modern standards the width and angle of the grooves in the pulleys are well standardized while requirements on belts are usually limited to that they should fit into the standardized pulley grooves.

11.3 Simple belt drives with two pulleys in the same plane

Most belt drives operate with two parallel shafts, each with a pulley and the pulleys located in the same plane, figure 11-2. The geometry of the belt is then determined by diameters d_1 and d_2 of the pulleys, belt length L and shaft centre distance c. Index 1 is used for the driving pulley (usually the small pulley) and index 2 for the driven pulley (usually the large pulley). Diameters of pulleys and lengths of belts can either be given in the pitch, datum or effective system. The arcs of contact between the belt and pulleys are given by arc of contact angle θ_1 and θ_2, the sum of which is 2π. Of the six geometrical quantities mentioned d_1, d_2, θ_1, θ_2, L and c, three can be given arbitrary values, after which the remaining three can be calculated with simple geometry. Assuming large shaft centre distance, i e:

$$c \gg \frac{d_1+d_2}{2} \qquad \qquad (11:4)$$

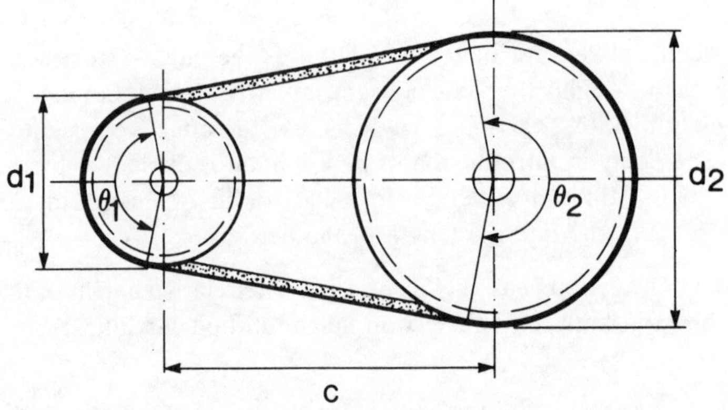

Figure 11-2. Basic geometry for V-belt drive with two pulleys in the same plane

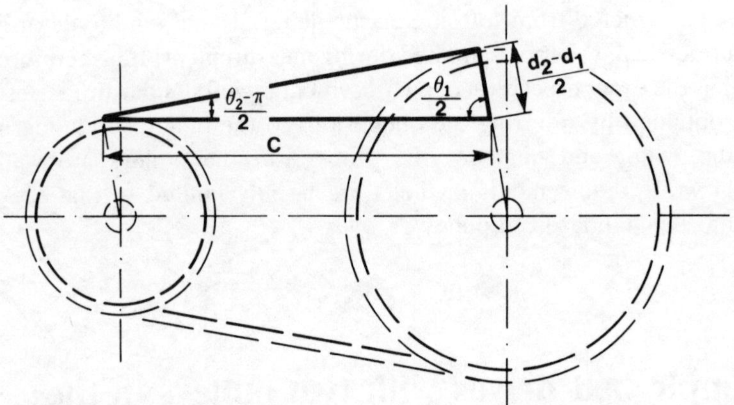

Figure 11-3. Geometrical calculation in V-belt drive with two pulleys

the approximations below can be used. Furthermore the bending stiffness of the belts, which tends to lift the belts between pulleys and reduce arc of contact somewhat, is neglected.

The arc of contact angle is calculated according to:

$$\cos\frac{\theta_1}{2}=\frac{d_2-d_1}{2c} \tag{11:5}$$

With an error not exceeding 1° for θ_1 greater than 140° the following approximation can be used:

$$\theta_1=\pi-\frac{d_2-d_1}{c} \tag{11:6}$$

150

Belt length L is calculated according to:

$$L = 2c \sin\frac{\theta_1}{2} + \frac{d_1\theta_1}{2} + \frac{d_2\theta_2}{2} \quad\dots\dots\dots (11{:}7)$$

After simplification is obtained:

$$L \approx 2c + \frac{\pi}{2}(d_1+d_2) + \frac{(d_2-d_1)^2}{4c} \quad\dots\dots\dots (11{:}8)$$

This last equation is widely used in the design calculations of belt drives.

11.4 Geometrically complicated drives

In addition to the simple belt drive, containing two pulleys operating in the same plane, geometrically more complicated drives are also to be found. These may for instance include crossed drives, quarter-turn drives, cf figure 2-5, and drives with idler pulleys, cf figure 14-6. Other examples are belt drives with more than two pulleys.

For *crossed belt drives* belt length L can be calculated according to:

$$L = 2c + \frac{\pi}{2}(d_1+d_2) + \frac{(d_2+d_1)^2}{4c} \quad\dots\dots\dots (11{:}9)$$

It can be noted that the only difference between this equation and the equation for the simple case, equation 11:8, is the use of a plus instead of a minus sign in the last parenthesis.

For *quarter-turn drives,* as well as for other drives with pulleys with angular displacement, simple equations for calculation of belt length are not available. Use can be made of estimations, or more accurate, but complicated calculations, best performed with the aid of computers.

For *belt drives with more than two pulleys* belt length L can be calculated as the sum of the span lengths between pulleys $\triangle L_s$ and the contact lengths on pulleys $\triangle L_\theta$, i e:

$$L = \Sigma\,\triangle L_{si} + \Sigma\,\triangle L_{\theta i} \quad\dots\dots\dots (11{:}10)$$

Here "i" denotes the number of the respective pulley, cf figure 11-4.

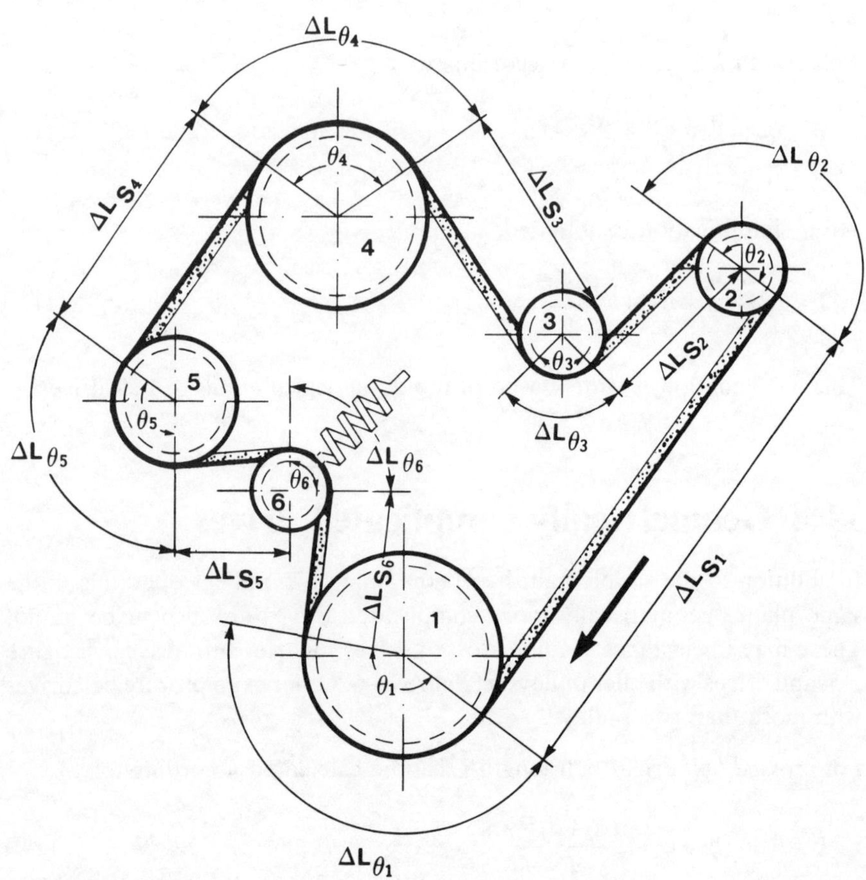

Figure 11-4. Belt drive with several pulleys (same example as in figure 2-4) with arcs of contact angle θ, span lengths $\triangle L_s$ and contact lengths at pulleys $\triangle L_\theta$ indicated

Span length $\triangle L_{si}$ is calculated according to, cf figure 11-3:

$$\triangle L_{si} = \sqrt{c_i^2 - \frac{(d_{i+1} - d_i)^2}{4}} \approx c_i \left[1 - 0,125 \frac{(d_{i+1} - d_i)^2}{c_i^2} \right] \quad \dots\dots\dots\dots\dots (11{:}11)$$

The contact length of the belt at pulleys $\triangle L_{\theta i}$ is calculated according to:

$$\triangle L_{\theta i} = \frac{\theta_i d_i}{2} \quad \dots\dots\dots\dots\dots\dots\dots\dots\dots\dots\dots\dots\dots\dots\dots\dots\dots (11{:}12)$$

For the calculation the coordinates of the shafts, alternatively shaft centre distances, and diameters of the pulleys are given. If coordinates for the shafts are given shaft centre distances can be calculated as the hypotenuse applying

152

the Pythagorean theorem. From a scaled drawing of the drive span length $\triangle L_{si}$ can be measured or calculated with the aid of equation 11:11. From the drawing arc of contact angles θ_i can be estimated and contact lengths at pulleys $\triangle L_{\theta i}$ can then be calculated with equation 11:12. Most convenient and also leading to the best accuracy, is to make the calculation with the aid of computer programs which are often put at the disposal of customers by belt manufacturers. The computer programs usually allow inside as well as outside placement of pulleys and in advanced calculation programs also make calculations possible for pulleys with angular displacement.

11.5 Synchronous drives

In geometrical calculations of synchronous drives it must be observed that belt pitch length L_p and the pulley pitch circumferences C_{pl} and C_{p2} are all integers of the pitch p_b, cf figure 3-21, i e:

$$L_p = z_b \cdot p_b \dotfill (11:13)$$

$$C_{pl} = z_1 \cdot p_b \dotfill (11:14)$$

$$C_{p2} = z_2 \cdot p_b \dotfill (11:15)$$

The integers z_b, z_1 and z_2 are the numbers of the cogs in the belt, and the numbers of the recesses in the small and large pulley, respectively. By the use of reinforcement of high modulus and by placing it as near the pitch zone as possible, i e at the root of the tooth, pitch will be almost constant during service, independent of the bending and power transmission of the belt.

12. Transmission of forces between belt and pulley

12.1 Introduction

For a belt drive to function the forces necessary for power transmission must be carried over from pulleys to belts. As mentioned in section 2.4 and illustrated in figure 2-7 force can either be transmitted by friction or by positive engagement and mechanical interlocking. Mechanical interlocking is a special case, found with cogged belts in synchronous drives, while all other types of belts rely on friction for their power transmission. In this chapter forces acting between belts and pulleys, for the transmission of power with the aid of friction, will be discussed.

The friction force F is calculated in the usual manner as the product of normal force F_N and coefficient of friction μ:

$$F = \mu \cdot F_N \quad\text{...} (12:1)$$

In utilizing the wedge effect, figure 12-1, normal force F_N, which is calculated from belt tension, will be enlarged compared to the vertical force on the belt side F_V depending on the value of belt or groove angle α according to:

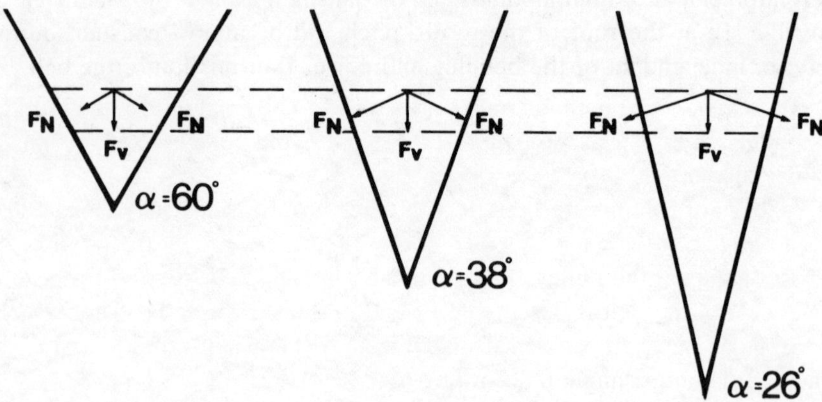

Figure 12-1. At a given vertical force F_V the normal force F_N at pulley sides will increase with decrease in groove angle α

$$F_N = \frac{F_V}{2\sin\frac{\alpha}{2}} \qquad\dotfill (12{:}2)$$

In design calculations use is often made of the apparent coefficient of friction μ_s defined as:

$$\mu_s = \frac{\mu}{\sin\frac{\alpha}{2}} \qquad\dotfill (12{:}3)$$

The friction force F can then be calculated starting from the vertical force F_V, which is calculated from belt tension, as:

$$F = \mu_s \cdot F_V \qquad\dotfill (12{:}4)$$

With $\alpha = 38° \quad \mu_s \approx 3\mu$.

12.2 Simple theory for forces between belt and pulley

For the calculations the following simplifying assumptions are made:

- Coefficient of friction μ is constant over the contact surface and independent of varying conditions
- Tension, friction, normal and centrifugal forces can be represented by a resultant force in the pitch zone of the belt section
- Friction forces between belt and pulley are either released with resulting slip or not released with no slip occurring
- Any slip between belt and pulley is assumed to be tangential with exception for the zones where the belt enters or leaves the pulley, where radial slip may also occur
- Pulleys are assumed to be rigid while belts can be elastically elongated

Forces acting at the pulley are shown in figure 12-2. The tension force in the tight side of the belt is designated F_1 and in the slack side F_2. The restoring force F_0 on the shaft can be calculated from belt tension forces F_1 and F_2 and the arc of contact angle θ according to:

$$F_0 = \sqrt{F_1^2 + F_2^2 - 2F_1F_2 \cdot \cos\theta} \qquad\dotfill (12{:}5)$$

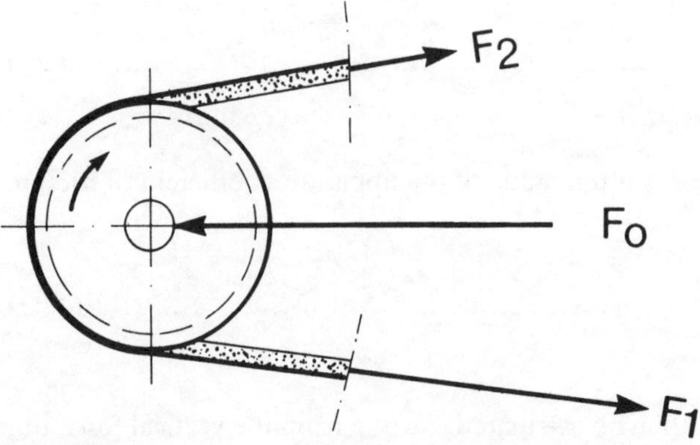

Figure 12-2. Forces acting at V-belt pulley

The bearing load F_0 can thus be described as the resultant in the parallelogram of forces where the sides are represented by the force vectors F_1 and F_2.

If pulleys are of approximately the same diameter and/or the shaft centre distance is large compared to pulley diameters, i e when belt strands form a small angle with the line connecting shaft centres, equation 12:5 can be simplified into:

$$F_0 \approx F_1 + F_2 \qquad (12:6)$$

For $\theta = 120°$ F_0 calculated according to equation 12:6 will be about 15% higher than if calculated according to equation 12:5. Bearing load F_0 calculated according to equations 12:5 and 12:6 represents the highest value obtained at a given belt pre-tension. It corresponds to the value obtained during standstill or at low belt speeds. At high belt speeds, i e when centrifugal forces are important and when belt tensioning at constant elongation is used, F_0 should be reduced by 2 F_c, where F_c is the centrifugal force projected along the length of the belt, cf equation 12:10 and figure 14-5.

The force F that transmits the power is the difference between the tight and slack side tension forces according to:

$$F = F_1 - F_2 \qquad (12:7)$$

As F_1 is larger than F_2 the tight side of the belt will be under greater strain (more elongated) than the slack side, cf section 17.2. The increase in strain from slack side to tight side and the decrease in strain form tight side to slack side will take place in the zones where the belt leaves the pulleys. The change

156

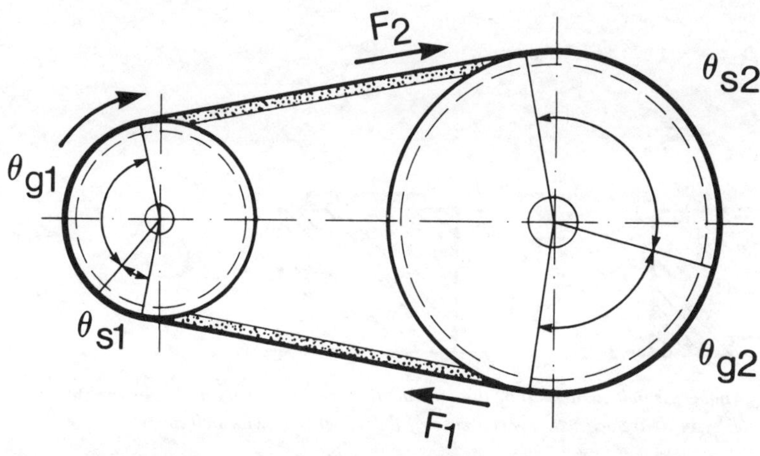

Figure 12-3. The arc of contact angle θ can be divided up into a zone with sliding θ$_g$ and a static zone without sliding θ$_s$

in length (strain) in these zones will lead to slip of the belt. The zone where slippage occurs, and thus where the friction forces are released, is termed slip zone, figure 12-3. The arc of contact angle θ can thus be divided into two zones, one θ_g where slip occurs and one θ_s (s for static) where no slip occurs:

$$\theta = \theta_g + \theta_s \quad\quad\quad\quad\quad\quad\quad\quad\quad (12:8)$$

In the two slip zones the increase and decrease in belt tension thus takes place, which is necessary for the force changes that cause power transmission according to equation 12:7. In the static zone no such change in belt tension takes place and consequently no contribution to force and power transmission is given.

For a belt running idle θ_g will approach 0, but it will increase with increased power transmission. When θ_g has reached θ the maximum power transmission capability of the drive has been achieved. Additional requirements for increased power from the driven unit will then lead to excessive slip. This will occur first at the pulley with the smallest arc of contact angle, i e the small pulley.

To derive the equations that relate F_1 and F_2 a small element in the belt is first considered. A flat belt can here be considered as a special case of a V-belt with a belt angle $\alpha = 180°$, figure 12-4. The element, according to figure 12-5, is then exposed to the following forces:

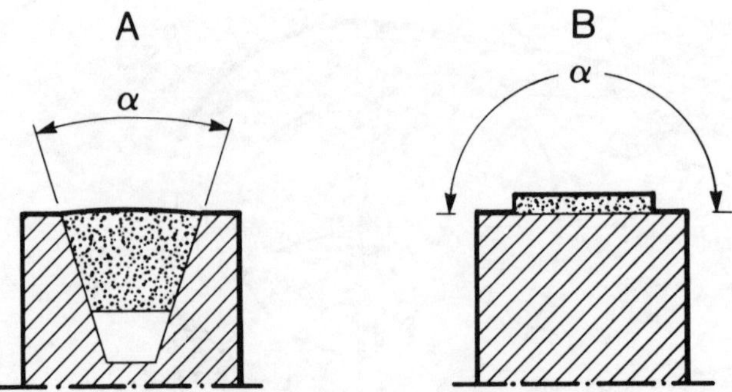

Figure 12-4. *Angles for belt element to be used in calculations for V-belt and flat belt respectively. The flat belt (B) can be treated as a special case of the V-belt (A) with belt angle α=180°*

- The belt tension force F with the differential increase dF
- The differential normal force dF_N (acting on each sidewall of the pulley groove)
- The differential frictional force μdF_N (acting on each sidewall of the pulley groove)
- The differential centrifugal force dF_{cr}

The requirement for equilibrium of forces in the radial direction then leads to:

$$dF_{cr} + 2\, dF_N \sin\frac{\alpha}{2} - (2F + dF)\,\frac{d\theta}{2} = 0 \quad\quad\quad (12{:}9)$$

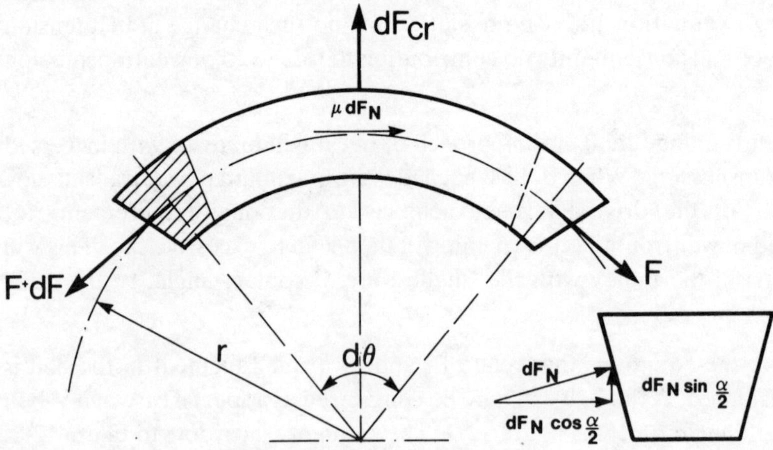

Figure 12-5. *Forces acting on a belt element in the sliding zone*

158

For a belt with belt section A, density ϱ and speed v the differential centrifugal force, cf figure 12-5, becomes:

$$dF_{cr} = A \varrho \, r \, d\theta \, \frac{v^2}{r}$$

which gives:

$$\frac{dF_{cr}}{d\theta} = A \varrho \, v^2$$

Now introduce F_c according to:

$$F_c = A \varrho \, v^2 \quad\dotfill\quad (12{:}10)$$

which can be taken as the projection of the centrifugal force along the length of the belt. By rearrangement and simplification, equation 12:9 becomes:

$$dF_N = \frac{(F - F_c) \, d\theta}{2} \cdot \frac{1}{\sin\frac{\alpha}{2}} \quad\dotfill\quad (12{:}11)$$

Force equilibrium for the tangential forces will give:

$$(F + dF - F) \cos\frac{d\theta}{2} - 2\mu \, dF_N = 0 \quad\dotfill\quad (12{:}12)$$

Using differentials $\cos\frac{d\theta}{2}$ can be approximated by 1. By elimination of dF_N the basic differential equation will be:

$$\frac{dF}{F - F_c} = d\theta \, \frac{\mu}{\sin\frac{\alpha}{2}} \quad\dotfill\quad (12{:}13)$$

Effective belt tension F_e is now defined as:

$$F_e = F - F_c \quad\dotfill\quad (12{:}14)$$

Combining equations 12:3, 12:13 and 12:14 leads to:

$$\frac{dF_e}{F_e} = \mu_s d\theta \quad\dotfill\quad (12{:}15)$$

This equation can now be integrated from the slack side from $\theta=0$ to $\theta=\theta_g$ leading to:

$$F_{e1}=F_{e2} \cdot e^{\mu_s \theta_g} \qquad\qquad\qquad (12{:}16)$$

This basic equation is widely used for engineering design calculations. In the German literature it is mostly called Eytelwein's or Euler's equation and in the English literature Reynold's equation.

Equation 12:16 is also valid for flat transmission belts if the apparent coefficient of friction, μ_s, is replaced by the coefficient of friction μ. For flat transmission belts of relatively large thickness and with the reinforcement close to the top side the difference in elongation obtained by changing from belt tension F_1 to F_2 and vice versa can largely be taken up by shear deformation in the rubber without slip occurring.

12.3 More advanced theory for forces between belt and pulley

The simple theory treated in section 12.2 above, leading to Eytelwein's equation 12:16, is adequate for most technical calculations. For more accurate calculations and for theoretical studies, aimed at the construction of new types of V-belts and increasing the understanding of mode of operation of V-belts, more advanced and complicated methods of calculation have been developed. These will, among others, take the following factors into account:

1. The dimensions of belt profile will diminish by lateral contraction during elongation of the belt and will narrow down when bending the belts over the pulleys, cf figure 5-3. These changes in profile will lead to the V-belt sinking down in the groove.
2. After the V-belt has entered the pulley groove, figure 12-6, it will gradually work itself deeper into the groove. The position of the V-belt in the pulley groove cannot then be described as circular, but is rather lika a helix. The gradual sinking into the groove also leads to radial slip usually not exceeding 1 mm. Due to the radial slip, the resultant frictional force will incline from the tangential direction. The magnitude of the inclination will depend on the ratio between the elastic stiffness along, and that transverse the direction of the belt. At high transversal stiffness the theory in section 12.2 is valid.

160

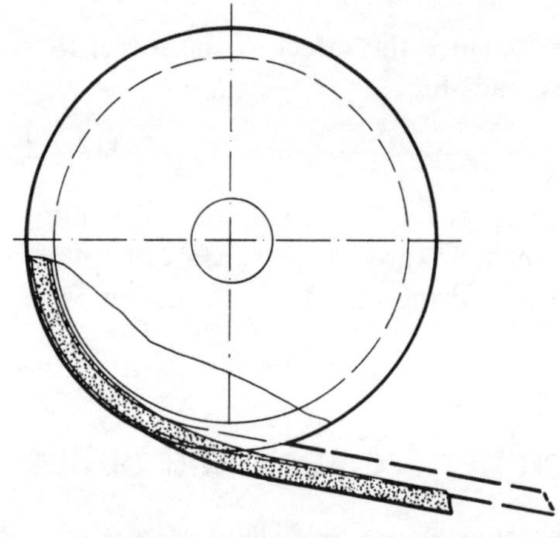

Figure 12-6. Radial movement of a V-belt after entering the pulley groove (continuous line) compared to an idealized, circular trace (dotted line)

3. Taking edge effects, bending moments on the belt section, etc, into account, one obtains uneven stress distribution on the contact surfaces between belt and pulley groove as well as on free surfaces of the belt section, figure 12-7. Stresses will usually increase towards the edges of the belt section. For more accurate calculations the uneven stress distribution must be taken into account and the simplifying assumption abandoned of representing all stresses by a force acting in the pitch zone of the belt profile and being evenly distributed over all contact and free surfaces.

One example of a case, where the simple theory discussed in section 12.2 is not sufficient, is the calculation of axial force F_a in a variator drive with split and axially adjustable pulley halves. The axial force is here used for

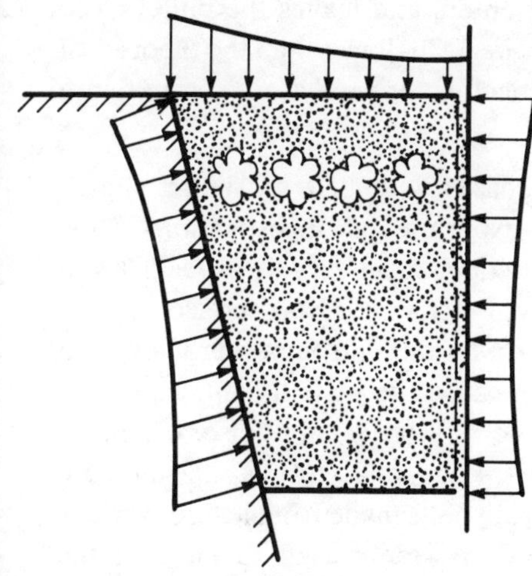

Figure 12-7. Typical compressive stress distribution in a V-belt section

adjustment of speed ratio. As it is to balance the effect of the forces that contribute to power transmission, this leads to:

$$F_a = k_1 (F_1 - F_2) \quad\dotfill\quad (12{:}17)$$

The value of the coefficient k_1 will be higher for the driving pulley than for the driven pulley and is also dependent on pulley groove angle α, arc of contact angle θ and coefficient of friction μ, cf also the discussion in section 14.8, equation 14:16.

12.4 Coefficient of friction between belt and pulley

The discussion above shows that the value of the coefficient of friction μ between belt and pulley is decisive for the ratio between belt tensions in the tight and slack side respectively, equation 12:16, and thus also for the force available for power transmission, equation 12:7. As pointed out in section 6.4.1, the coefficient of friction of polymeric materials is not a well-defined material property, but its value depends largely on testing conditions, such as contact force, speed of sliding, temperature, contact area and condition of surfaces in contact. To obtain data, relevant for belt drives, belts must thus be manufactured from materials, the frictional properties of which are to be studied, when mounted and tested between a driving and driven pulley. The power transmitted as a function of slip is determined, whereafter the coefficient of friction can be calculated with equation 12:16. The calculations will lead to a type of average coefficient of friction over the whole arc of contact. The values thus obtained are often unexpectedly low depending on disturbances in the zones where the belt enters and leaves the pulley. The "effective" arc of contact angle will usually be 5–30° lower than the theoretical (geometrical) one. Some data from practical experience are given in figure 12-8.

The coefficient of friction normally increases with belt or sliding speed. Suggested equations and results from experimental investigations show roughly a doubling of the coefficient of friction, from very low speeds up to high speeds. The coefficient of friction may also change during service. Ageing and wear processes make the rubber surface harder and smoother leading to a lowered coefficient of friction. This can cause the belts to slip, especially when starting up, which is often noticeable as a squeaking noise. Chemical attack or swelling by liquids may soften the surface, make it sticky and increase its coefficient of friction. For flat transmission belts made from leather various resins, so-called belt dressings, were earlier widely used to increase the

Application	Contacting material	Coefficient of friction μ
V-belt drive	Styrene, natural or chloroprene rubber on steel	0,20–0,30
V-belt drive	Special urethane rubber on steel	0,40–0,50
Flat belt drive	Rubber on steel	0,30–0,40
Flat belt drive	Chrome leather on steel	0,70–0,80
Car tyre on dry road	Rubber on asphalt/concrete	0,6–0,8
Car tyre on wet road	Rubber on wet asphalt/concrete	0,3–0,4
Conveyor belt, dry conditions	Rubber on steel	0,35–0,40
Conveyor belt, wet conditions	Wet rubber on steel	0,10–0,20
Conveying of materials	Particulate minerals on rubber	0,25–0,40

Figure 12-8. Typical values of coefficient of friction in some practical applications

coefficient of friction. With modern belt materials this method has lost its importance and is discouraged for V-belts, cf section 9.6.

The most obvious way of increasing friction in a V-belt drive is by lowering the belt angle, equation 12:3. Too small belt angles α will, however, lead to locking of the V-belts in the pulley grooves and may cause their rupture.

12.5 Slip

The discussion in section 12.2 shows that the slip zone θ_g increases with an increase in power transmission. When θ_g has reached θ, slip will occur over the whole arc of contact, and maximum power transmission capability has been reached. Attempts to additionally increase power transmission will only lead to excessive slippage, cf also sections 16.5 and 17.2. This will result in unacceptable wear and heating of the belts and to shortened belt life. Excessive heating may lead to chemical and/or thermal breakdown, which in extreme cases transforms the rubber material to a semi-solid state (smell of burnt rubber). This may "lubricate" the surface and additionally decrease power transmission capability. The pulleys and bearings, expecially at the small pulley, may also be overheated as a result of the slippage. Excessive slip should thus be avoided by proper design of the belt drive, avoiding over-loads and using the correct belt tension, cf chapter 14.

13. Forces and stresses in V-belts

13.1 Introduction

In this chapter stresses and strains in V-belts will be discussed. The discussion can to a large part also be applied to other types of belts. The treatment is partly based on the discussion of forces between belts and pulleys in chapter 12. Firstly the various types of stresses will be described, after which their accumulative effect will be considered. The discussion is complicated by the fact that stresses not only vary over the belt section area but also along the length of the belt. Simplifying assumptions will be made. More detailed and more accurate treatments are to be found in scientific literature.

13.2 Tensile stresses due to belt pre-tension and power transmission

According to section 12.2 a belt transmitting power by friction must be under pre-tension to generate the necessary friction forces. Power transmission is proportional to the difference in belt tension between the tight side F_1 and the slack side F_2 of the belt, equation 12:7. With increasing belt speed the effects of centrifugal forces on the belt become increasingly important. These strive to lift the belt from the pulley groove and thereby lead to a reduction in normal force. The total belt tension F must thus be reduced by the centrifugal tension force F_c in order to get the effective belt tension F_e according to equation 12:14. Introducing belt section area A, the effective belt tension stress difference σ_e can be calculated as:

$$\sigma_e = \frac{F_e}{A} \quad\dotfill\quad (13:1)$$

This way of describing stresses is often used, even though it assumes a homogeneous material, which is obviously not at hand in V-belts with reinforcement of much higher stiffness than the surrounding rubber. Belt tension forces are mainly taken up by the reinforcement and therefore lead to very small stresses and strains in the rubber material.

13.3 Tensile stresses caused by centrifugal forces

When belts pass around pulleys they will be subjected to centrifugal forces, mainly depending on belt speed, which try to lift the belt out of the pulley groove, figure 13-1. Centrifugal forces are undesirable as they will increase tensile stresses in the belts, while at the same time reducing normal forces.

The centrifugal force is the most important factor in determining limits for belt speed and thus to a large extent power transmission capability, cf section 16.6.

For a belt of section area A and density ϱ that runs with belt speed v or angular speed ω over a pulley of diameter d the centrifugal force F_c projected along the belt can be calculated, cf equation 12:10, according to:

$$F_c = A \varrho \left(\frac{d\omega}{2}\right)^2 = A \varrho v^2 \quad\quad\quad (13:2)$$

The centrifugal force F_c gives rise to a centrifugal tensile stress σ_c in the belt calculated as:

$$\sigma_c = \frac{F_c}{A} = \varrho\left(\frac{d\omega}{2}\right)^2 = \varrho v^2 \quad\quad\quad (13:3)$$

Centrifugal tensile stress σ_c will thus only depend on density ϱ and belt speed v, but it is independent of the diameter of the pulleys. Take as an example $\varrho = 1\ 300\ kg/m^3$ for a classical V-belt with A section and a section area $A = 0,82$

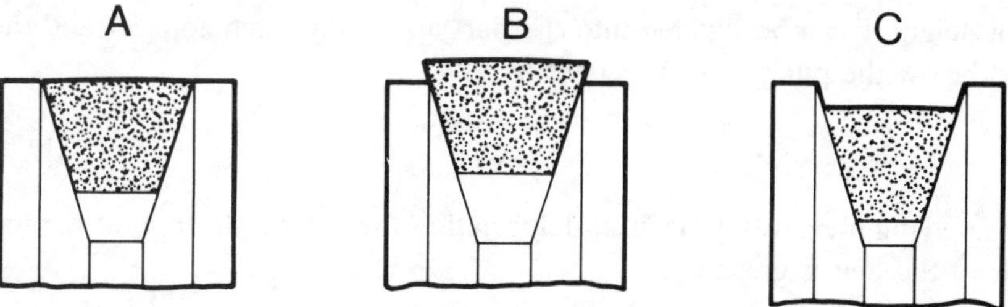

Figure 13-1. The effect of centrifugal force on the position of the V-belt in the groove at normal (A), high (B) and low (C) belt speed

cm^2 equation 13:2 will give F_c=11 N for v=10 m/s and F_c=267 N for v=50 m/s. The example illustrates how fast centrifugal forces will increase with an increase in belt speed.

13.4 Bending (flexural) stresses

During the bending of the belt over the pulley flexural stresses are generated in the form of tensile stresses in the upper part of the belt and compressive stresses in the lower part, figure 13-2. These flexural stresses are superimposed upon other tensile stresses in the belt and can lead to large local increases in tensile stress, without, however, affecting the average tensile stress level.

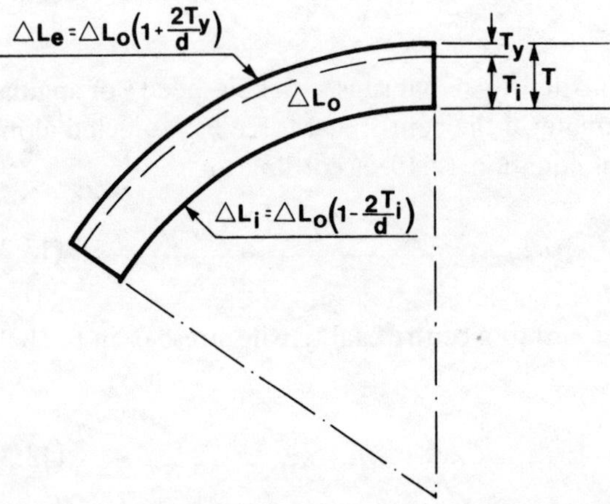

Figure 13-2. Deformation of belt during bending. The dotted line indicates the pitch zone where the length of the belt element is $\triangle L_0$. With application of Hooke's law the outer length (effective length) of the belt element $\triangle L_e$ and its inner length $\triangle L_i$ are calculated according to the figure

Belt height T can be divided into the part above the pitch zone T_y and the part below the pitch zone T_i according to:

$$T=T_y+T_i \qquad (13:4)$$

By applying elementary flexural deformation theory the strain ε_y at the top side of the belt is given by:

$$\varepsilon_y=\frac{2T_y}{d} \qquad (13:5)$$

166

The compressive strain ε_i, as a consequence of the bending on the underside of the belt, is then given by:

$$\varepsilon_i = \frac{2T_i}{d} \quad \dots\dots\dots\dots\dots\dots\dots\dots\dots\dots\dots\dots\dots\dots\dots\dots\dots\dots\dots\dots\dots(13:6)$$

A simplifying assumption, often applied in design calculations, is that the belt is made of a homogeneous material of Young's modulus E, whereby $T_y = T_i = \frac{T}{2}$. By applying Hooke's law the tensile stress due to flexure σ_b can then be calculated as:

$$\sigma_b = \frac{E\ T}{d} \quad \dots\dots\dots\dots\dots\dots\dots\dots\dots\dots\dots\dots\dots\dots\dots\dots\dots\dots\dots\dots (13:7)$$

The largest tensile stresses due to bending are thus to be expected in belts with a large height T running over pulleys with small diameters d.

The elongation or compression at the surfaces of the V-belt due to bending is usually a few per cent, exceptionally up to 10%. The resulting bending stresses will be low due to the low modulus of elasticity of rubber. Higher flexural stresses may be noted in the reinforcement due to its higher stiffness. These are, however, difficult to calculate as their magnitude will depend on the unknown relative internal mobility of the individual fibres.

13.5 Other types of stresses

Among other stresses that may appear in V-belts the following ones should be mentioned:

Torsional stresses will appear in belts with angular displacements, e g quarter-turn and crossed belt drives, cf figure 2-5.

Compressive stresses in the contact zone between belts and pulley groove side-walls, cf figure 12-7.

Shear stresses in the transmission of friction forces from pulleys to the reinforcement. The tension force in the belt is mainly taken up by the reinforcement and only to a small degree by the rubber material. This means that large forces in the form of shear stresses have to be transmitted from the pulley to the reinforcement. The development of improved reinforcement materials, which can take higher tensile forces and transmit more power, increases the demand on the ability of the rubber material to withstand high shear stresses. The largest shear stresses will be reached where the shortest distance between reinforcing cords and pulley groove sidewalls exist, i e at the upper corner of the V-belt.

13.6 Total stress

The total stress in a belt σ_t can now be calculated as:

$$\sigma_t = \sigma_e + \sigma_c + \sigma_b = \sigma_e + \varrho v^2 + \frac{E\,T}{d} \qquad\qquad (13:8)$$

The total stress is not constant but will vary over the section area of the belt as well as along its length, figure 13-3. The effective belt tension stress σ_e is here divided into two parts, one σ_{e2} representing slack side tension to which is added σ_{e1} in the tight side. First looking at the variation over belt section area, it can be noted that the largest stress is obtained at the top side of the belt when the tight side passes over the small pulley. Obviously, the total stress is larger in the tight side than in the slack side of the belt, and gets an extra contribution when the belt is bent over pulleys.

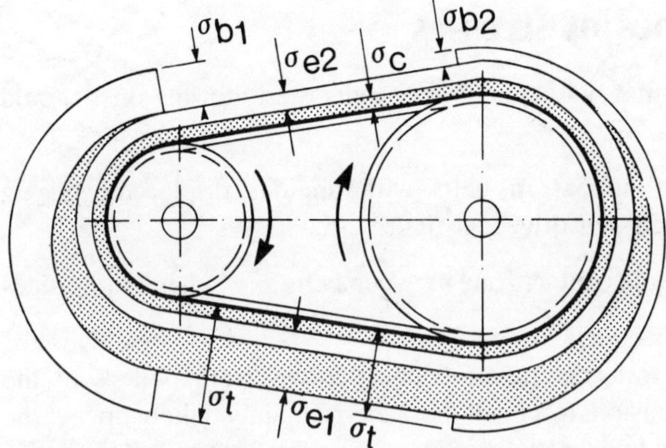

Figure 13-3. Variation of tensile stresses in a V-belt during service around its circumference

14. Tensioning of V-belts

14.1 Introduction

The importance of applying proper belt pre-tension in belts operating with power transmission by frictional forces is pointed out in sections 9.3 and 12.2. Too low belt tension will lead to excessive slip of the belt, frictional heating leading to high temperatures, catastrophic wear and reduced power transmission. Too high belt tension will lead to shortened belt life, increased loads on shafts and bearings and decreased mechanical efficiency of the drive. Proper belt tension during service is obtained, by the selection of a proper pre-tension before starting the drive. In this chapter various methods to select, measure, adjust and maintain proper belt tension will be discussed. The effective use of proper pre-tension assumes that the belt drive is designed correctly.

14.2 Selection of proper pre-tension

14.2.1 Visual inspection

The lowest possible belt tension that doesn't lead to excessive slip is usually considered to be the best choice. The selection of a proper level can therefore be done by observing actual slip during operation of the belt drive. If excessive slip is noted, then pre-tension should be increased. Inspection can be made visually or alternatively the rotation speed of the shafts can be measured, e g with the aid of a tachymeter, and the slip then calculated. It should preferably not exceed a few percent. Visual inspection of the slack side of the belt in operation can reveal if belt tension is too high. The slack side should preferably not be straight, but form a catenary curve, which may be observed visually, at least if shaft centre line distance is not too small. If the slack side of the belt is completely straight, then the belt tension is probably too high.

14.2.2 Measurement and calculation of belt tension with the aid of a tensiometer

The adjustment and measurement of pre-tension with the aid of a tensiometer is discussed in section 9.3.3. This is the most common method of measurement used, often referred to in standards and in V-belt manufacturers' catalogues. In this simple instrument the force required to deflect the belt at the middle of its span a distance δ, usually $\dfrac{1}{64}$ of the free span length, is determined. In belt drives with not too short shaft centre distance and pulleys not differing too much in diameter, span length can be approximated with shaft centre distance c. The relation between measured force F_m and belt tension force F, cf figure 14-1, can then be calculated as follows:

$$\delta=\frac{c}{64} \qquad (14:1)$$

If the angle between the deflected belt and the original belt line is called ß, cf figure 14-1, then the following expressions will be valid:

$$F_m=2F \sin ß \approx 2ß\,F \qquad (14:2)$$

$$ß=\arctan\left(\frac{2\delta}{c}\right)\approx\frac{2\delta}{c} \qquad (14:3)$$

$$F=\frac{F_m\,c}{4\delta} \qquad (14:4)$$

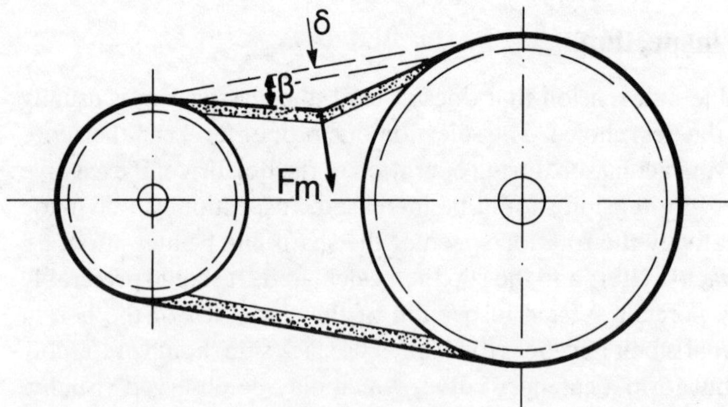

Figure 14-1. Measurement of belt tension F from the deflection δ under the influence of a measuring force F_m when the span length is approximated with shaft centre distance c

And by combining equations 14:1 and 14:4:

$$F = 16F_m \dots\dots\dots\dots\dots\dots\dots\dots\dots\dots\dots\dots\dots\dots\dots\dots\dots\dots \quad (14:5)$$

In catalogues from the V-belt manufacturers recommended tensiometer values leadning to correct belt tension are usually given, cf figure 9-10. Sometimes also maximum and minimum values for tensiometer force F_m are given, e g calculated according to:

$$(F_m)_{max} = \frac{1,5F + k_0}{16} \dots\dots\dots\dots\dots\dots\dots\dots\dots\dots\dots\dots\dots\dots \quad (14:6)$$

$$(F_m)_{min} = \frac{F + k_0}{16} \dots\dots\dots\dots\dots\dots\dots\dots\dots\dots\dots\dots\dots\dots \quad (14:7)$$

The coefficient k_0 will depend on the size of the belt section. Tabulated data is available where this procedure is recommended. Optimum belt pre-tension, and thus tensiometer values, should be adapted to the power to be transmitted. The maximum power that can be transmitted will depend on the diameter and speed of rotation of the small pulley and the speed ratio used, as shown in chapter 16 and Appendix G. These factors should therefore be considered when making up tables for recommended values of pre-tension, cf figure 9-10.

14.2.3 Measurement of strain in the belt

The tensiometer is sometimes not considered practical or not accurate enough, e g for joined V-belts or heavy V-belt sections. Belt tension can then be calculated from the measured elongation of the belt when applying the pre-tension. The method assumes that the relation between belt tension force and belt elongation is known, alternatively that the Young's modulus E, or the spring constant in tension, of the belt is known and Hooke's law can be assumed to be valid. Where this method is recommended by V-belt manufacturers, their catalogues give data on Young's modulus, or the spring constant in tension, of various belts to make these calculations possible. Proper pre-tension of the belt usually corresponds to a belt elongation of 0,5–1,5%, cf figure 9-7.

14.2.4 Calculation from power to be transmitted

The pre-tension needed in a V-belt can also be calculated starting from the power that has to be transmitted by the belt drive. It is usually assumed that tight side tension force F_1 should be 5 times the slack side tension force F_2. By applying equation 16:2 this will lead to:

$$P=(F_1-F_2)\ v=(5F_2-F_2)\ v=4F_2\ v \dots\dots\dots\dots\dots\dots\dots\dots\dots\dots\dots(14{:}8)$$

The pre-tension force F is approximately half the reaction force F_0 on the shafts, so by applying equation 12:6:

$$F\approx\frac{F_0}{2}\approx\frac{F_1+F_2}{2}\approx\frac{5F_2+F_2}{2}\approx3F_2 \dots\dots\dots\dots\dots\dots\dots\dots (14{:}9)$$

Combining equations 14:8 and 14:9:

$$F\approx3F_2\approx\frac{3P}{4v} \dots\dots\dots\dots\dots\dots\dots\dots\dots\dots\dots\dots\dots (14{:}10)$$

This simplified calculation neither takes centrifugal forces nor arc of contact angles different from 180° into account. With regard to these factors more accurate equations are presented in many V-belt catalogues, usually of the following type:

$$F=k_1\frac{2{,}5-k_3}{k_3}\cdot\frac{P}{Zv}+k_2\ v^2 \dots\dots\dots\dots\dots\dots\dots\dots\dots (14{:}11)$$

Coefficients k_1 and k_2 are given in tables, where this method of calculation is recommended while Z is the number of belts in a multiple V-belt drive. The coefficient k_3 is identical to the correction factor for arc of contact angle, see section 18.9.4 and figure 18-8. The values for belt pre-tension force calculated can then be adjusted and checked with the method described in sections 14.2.2 and 14.2.3.

14.3 Belt tensioning by stretching (constant elongation)

This method utilizes the elasticity of the belt. The shaft centre distance c, calculated from the unstretched length of the belt, is increased by an amount

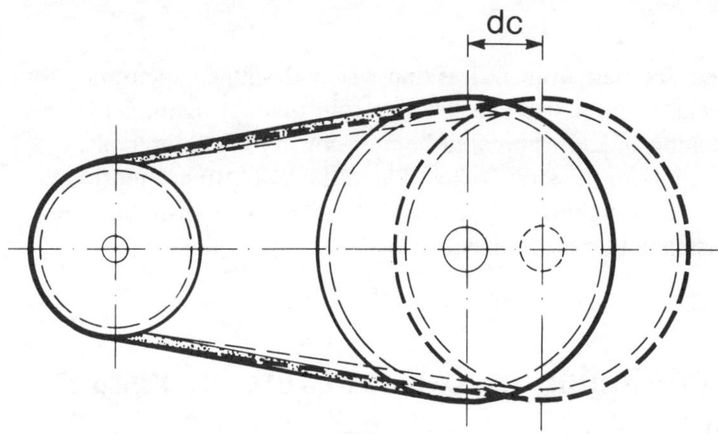

Figure 14-2. Belt pre-tensioning with constant elongation (fixed shaft centre distance)

dc, figure 14-2, until the necessary pre-tension force is reached. Stretching can be performed with the aid of adjusting screws, slide rails or other mechanical devices, figure 14-3. As the adjusting screws stay in a locked position they will keep the belt at a fixed elongation. This method of adjusting belt tension is probably the most widely used one, thanks to the simple and cheap devices

Figure 14-3. Pre-tensioning by constant elongation using adjusting screws and slide rails on the motor mounting

that may be used. Its main drawback is that it includes no compensation for the continued relaxation of stress in the belt with time, cf figure 6-14. This will require checking and adjustment of belt tension at regular intervals. One way of compensating for the stress relaxation in the belt is to pre-tension the belt to a belt tension higher than required for transmission of the power involved, for instance to a 50% higher value.

14.4 Belt tensioning by constant force (constant load)

With this method one of the shafts is freely movable along the line connecting the two shafts, figure 14-4. With the aid of various devices, e g weights or springs, the shaft is given a constant reaction force F_0, to tension the belt. This device, as well as the device mentioned in section 14.3, will operate equally well in both directions of rotation. Creep (increase in belt length during service) will not affect belt tension, making readjustment unnecessary.

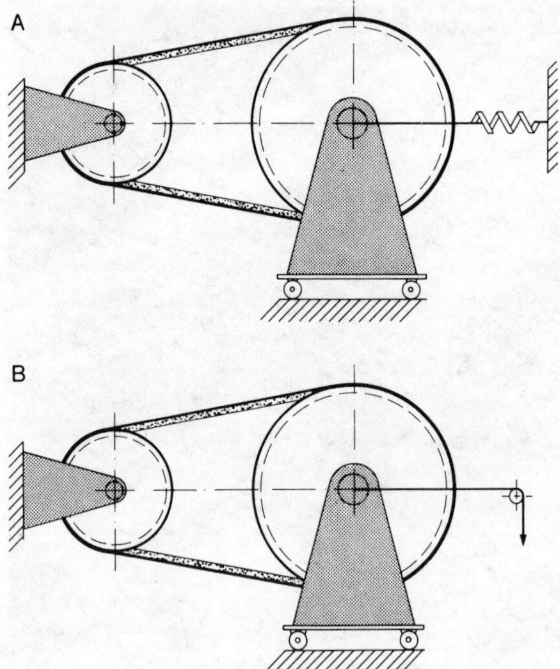

Figure 14-4. Belt pre-tensioning by constant load from a spring (A) or weight (B)

Belt pre-tensioning Factors affected	Constant elongation	Constant load
Shaft and bearing load	Decreases	Constant
Sum of tight and slack side belt tension force	Constant	Increases
Tight side belt tension force F_1	Decreases	Increases
Power rating	Increases to a maximum and then decreases	Increases up to rupture

Figure 14-5. Comparison of how different factors are affected by variation in belt speed and thereby centrifugal forces at pre-tensioning with constant elongation or constant force, respectively

The most important principal difference between belt tensioning with constant elongation and constant load is the effect of the centrifugal forces which is best noticeable at high belt speeds. The mathematical analysis will not be carried through here but its results can be summarized as in figure 14-5. At fixed elongation the centrifugal forces strive to throw the belt off the pulleys which decreases bearing loads and tight side tension F_1 reducing power transmission. When tensioning with constant load a different behaviour is observed, cf figure 14-5. As tensioning with constant elongation is by far the most widely used method most statements made in the literature on the effect of increasing belt speed and thereby increasing centrifugal forces, relate to this method even though it is not always clearly pointed out. It should be noted that the discussion behind figure 14-5 is only concerned with conditions in the belt drive and does not take into account the fact that centrifugal forces will reduce the effective belt tension, available for power transmission, assuming a maximum allowable belt tension with reference to the strength of the belt. This effect on effective belt tension is also valid for pre-tensioning at constant elongation as it is for pre-tensioning at constant load. This will limit the power that can be transmitted with increasing belt speed, cf section 16.6.

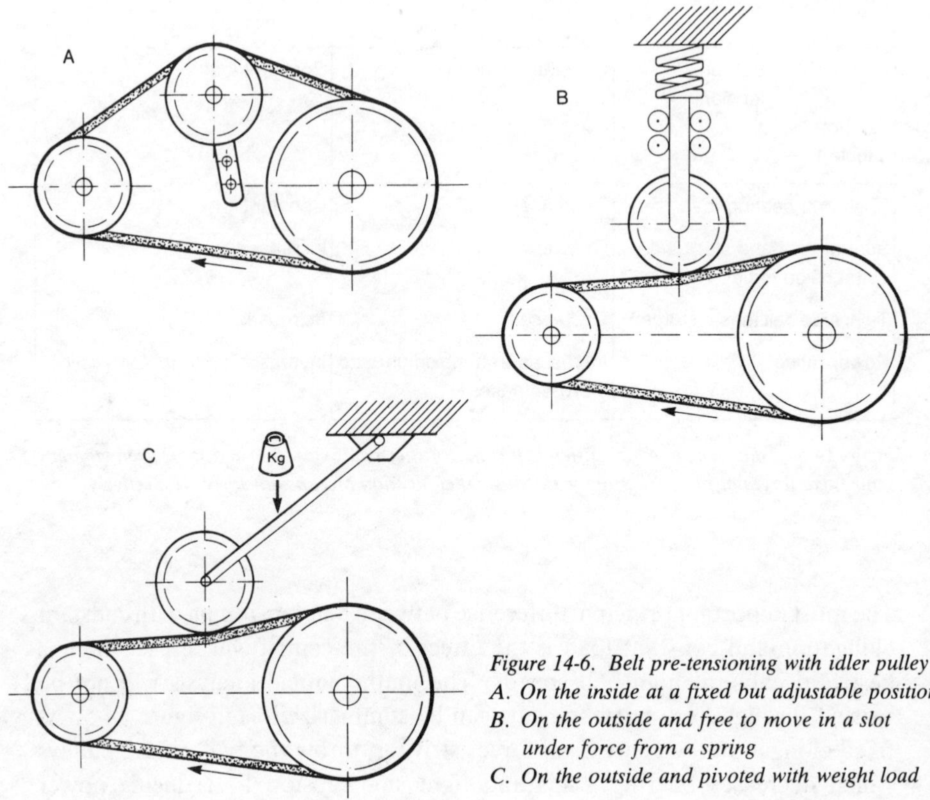

Figure 14-6. Belt pre-tensioning with idler pulley
A. On the inside at a fixed but adjustable position
B. On the outside and free to move in a slot
under force from a spring
C. On the outside and pivoted with weight load

14.5 Belt tensioning with idler pulleys

One way of tensioning V-belts is to use a flat or grooved pulley, rotating with the belt and pressed against it, figure 14-6. The method has the advantage that both driving and driven shafts can be stationary, which is especially of advantage in large and heavy machines. A drawback is that the belt will be exposed to extra bending stresses and thereby to increased fatigue strains and shortened belt life. Fatigue stresses will be larger if the idler pulley is placed on the outside of the belt (double-bending occurs) than on its inside (double-bending is avoided) and larger if it is placed on the tight side of the belt rather than on the slack side. With regard to fatigue the following placement of idler pulleys are then possible (in order of preference):

1. On the inside of the slack side of the belt
2. On the inside of the tight side of the belt

3. On the outside of the slack side of the belt
4. On the outside of the tight side of the belt

Belt life with a placement according to alternative 1 can be up to 10 times larger than with a placement according to alternative 2. For flat belts or hexagonal belts double-bending is not so critical, and so the choice of placement on the inside or outside of the belt becomes of less importance. In some special types of V-belts the reinforcement has been located deeper below the top surface so as to better accept double-bending at idler pulleys, cf figure 3-25 B. This advantage has to be paid for by accepting lower power ratings.

Idler pulleys should preferably be placed at one third the span length from the driving pulley on the slack side of the belt and at one third the span length from the driven pulley on the tight side of the belt. This placement will minimize vibrations in the belt and keep noise level down. By placing the idler pulley on the outside of the belt close to the small pulley, its arc of contact angle will be increased and thereby the power transmission capability of the drive. At the same time belt life may be reduced due to the double-bending.

Idler pulleys may have either normal V-belt grooves or a flat cylindrical surface. The flat pulleys are the only ones that can be used on the outside of the belts. Diameters of idler pulleys should not be too small. Examples of recommended minimum diameters are found in figure 14-7. As a rule, inside idler pulleys should have a diameter at least corresponding to the smallest

Profile	Smallest recommended pulley diameter according to RMA IP-21, mm	Smallest recommended diameter of internal, grooved idler pulley, mm	Smallest recommended diameter of internal, flat idler pulley, mm	Smallest recommended diameter of external idler pulley, mm	Smallest recommended width of flat idler pulley, mm
HAA	80	70	57	108	25
HBB	145	102	95	152	32
HCC	235	146	121	216	38

Figure 14-7. Minimum recommended diameter and width of idler pulleys for agricultural drives with hexagonal belts according to ASAE S211.3 (for additional information on hexagonal belts, see figure 3-13). It should be noted that this standard permits smaller diameters of the idler pulleys than normally recommended for minimum pulley diameters according to RMA IP-21, probably due to the fact that the RMA standard refers to industrial belts and the ASAE standard to agricultural belts with considerably lower requirements on belt life

Figure 14-8. Spring loaded idler pulley placed on the outside for adjustment of belt tension

pulley diameter recommended for the belt section in question and outside idler pulleys a diameter of at least 1,3–1,5 times this value.

Idler pulleys may either be mounted in fixed, but adjustable positions or be movable in grooves or by pivoting and exerting a constant force from a spring or weight loading, cf example in figure 14-8. With the use of a fixed but adjustable position conditions equivalent to the pre-tensioning of belts by stretching are obtained, cf section 14.3. Belt tension will then gradually decrease with time due to creep and stress relaxation of the belts. Checking and readjustment of belt tension will then be required. When the idler pulley exerts a constant force (gravitational force or spring loading) conditions similar to those described in section 14.4 are obtained and no readjustment of belt tension will be required. One important difference is that the idler pulley will only govern belt tension in the slack side while tight side tension can increase in proportion to the power transmitted. In this way the belts will not be loaded more than what is required for the power to be transmitted, which will have a positive influence on the belt life.

Idler pulleys are sometimes used for the disengagement of the drive, i e by releasing the force on the belt from the idler pulley the belt will slip totally and no power will be transmitted. Idler pulleys will also increase the shock absorption capability of the belt drive. Design calculations in connection with the use of idler pulleys are discussed in section 19.3.

14.6 Belt tensioning by the weight of the driving unit

In belt drives with strands running vertically belt tensioning can be obtained by utilization of the mass m of the driving unit pivoted with lever length l_0, figure 14-9. If effective belt tension forces are designated F_{e1} and F_{e2} operating with lever lengths l_1 and l_2 respectively moment equilibrium will give:

$$F_{e1}\,l_1 + F_{e2}\,l_2 = mg\,l_0 \hspace{2cm} (14{:}12)$$

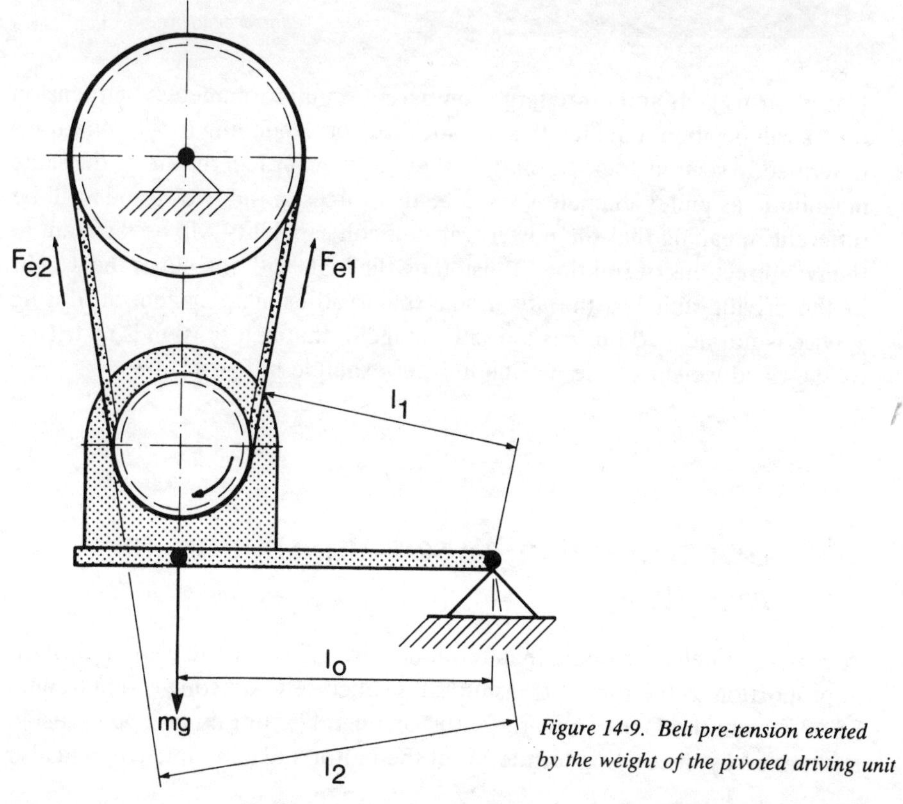

Figure 14-9. Belt pre-tension exerted by the weight of the pivoted driving unit

179

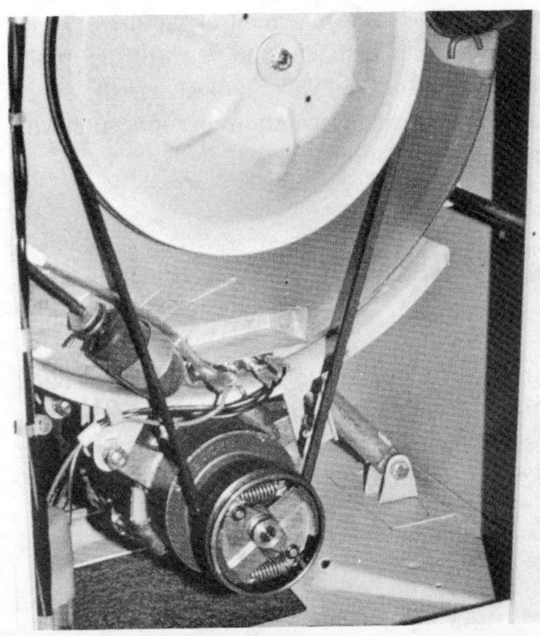

Figure 14-10. The weight of the pivoted electrical motor on a washing machine gives the required belt tension

If lever arms l_0, l_1 and l_2 are large compared to pulley diameters belt tension forces will be about equal in the two strands, corresponding to the conditions described in section 14.4. If, on the other hand, lever lengths are of the same magnitude as pulley diameters the effective forces in the two strands will be different, meaning that the power transmission capability will be different in the two directions of rotation. Tensioning the belts with the aid of the weight of the driving unit has the advantage that readjustment of tension during service is not needed but has a disadvantage in that belt tension is restricted by the fixed weight of the driving unit, cf example in figure 14-10.

14.7 Belt tensioning by self-adjusting motor mounting

A device for belt tensioning that automatically adjusts the tension in the belt in proportion to the torque transmitted is called self-adjusting (often termed SESPA from the German word "selbstspannend"). In this case belt tension is obtained by the driving torque M_1 of the motor and to a limited extent also

180

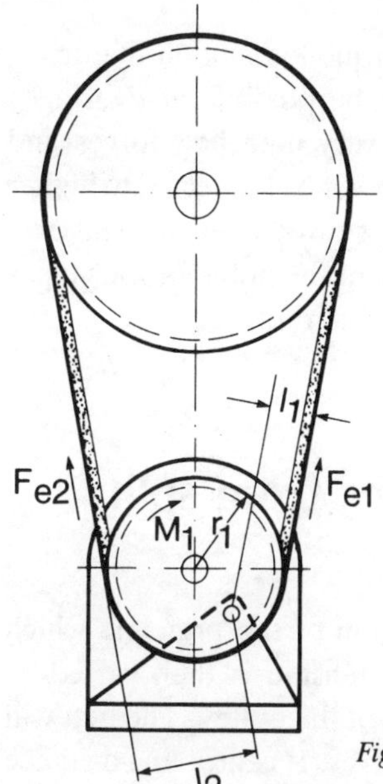

Figure 14-11. Belt pre-tensioning by self-adjusting motor mounting

by the weight of the driving unit. The pivoting shaft is placed between the two strands of the belt and eccentrically relative to the motor shaft, figure 14-11. If strands with effective belt forces F_{e1} and F_{e2} operate on lever arms with lengths l_1 and l_2 respectively the following equations can be set up:

$$F_{e1}\, l_1 = F_{e2}\, l_2 \qquad\qquad (14:13)$$

$$\frac{F_{e1}}{F_{e2}} = \frac{l_2}{l_1} \qquad\qquad (14:14)$$

The ratio between effective tight side tension F_{e1} and effective slack side tension F_{e2} will thus be determined by the ratio between lever arms l_1 and l_2. The difference between F_{e1} and F_{e2} will be related to the torque loading M_1 which with application of equations 16:1 and 16:3 can be given as:

$$M_1 = r_1\, (F_{e1} - F_{e2}) \qquad\qquad (14:15)$$

To reach the high ratio desired between F_{e1} and F_{e2}, it will thus be necessary to have l_1 small in comparison to l_2. If the direction of rotation is reversed, the belt will be repelled from the small pulley and fail to grip on it. With the correct direction of rotation effective belt force difference will increase according to equation 14:5 in proportion to the torque loading on the drive. Increased power transmission capability will thus be possible at overloads without risk for excessive slip. This can lead to very high belt forces and rupture in the belt. On the other hand the belt will never be subjected to higher belt tension than is required for transmitting the power needed. This will increase mechanical efficiency and the working life of the belt. Various types of self-adjusting belt tensioning devices are on the market.

14.8 Belt tensioning by axial force from split pulleys

Belt tensioning can also be performed by utilization of the principle which forms the basis for the belt variator drives. Belt tension is then varied by changing the axial force F_a between the two halves of the pulley. The belt will then change its radial position between the two halves. If centre line distance between the two shafts is locked, radial position will in large be maintained but belt tension changed. The relation between axial force F_a and effective belt tension force in the two strands F_{e1} and F_{e2} can then be expressed as, figure 14-12:

$$F_a = \int_0^{\theta_g} \cos\frac{\alpha}{2}\, dF_N \quad \dots\dots\dots\dots\dots\dots\dots\dots\dots\dots\dots\dots\dots\dots\dots\dots\dots\dots\dots\dots \quad (14:16)$$

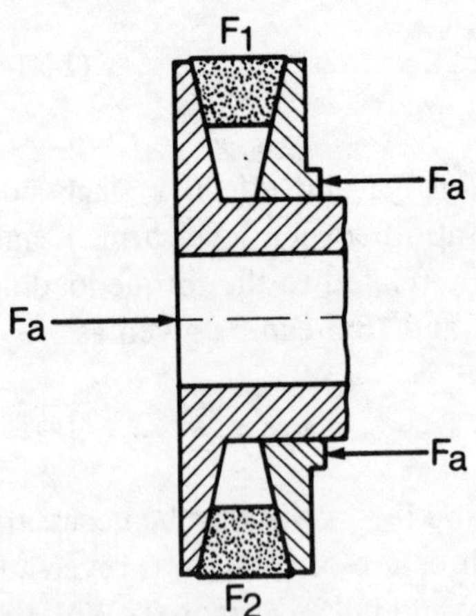

Figure 14-12. Belt tension forces F_1, and F_2 and axial force F_a at split variator drive pulley

This equation can be evaluated and arranged to:

$$F_a = \int_{F_{e2}}^{F_{e1}} \frac{\cos \frac{\alpha}{2}}{2\mu} \, dF_e = (F_{e1} - F_{e2}) \frac{\cos \frac{\alpha}{2}}{2\mu} \quad \dots\dots\dots\dots\dots\dots\dots\dots\dots\dots (14{:}17)$$

This simplified calculation assumes that the total arc of contact is utilized for power transmission, i e $\theta_g = \theta$. More accurate calculations require that also the static zone of the arc of contact and radial slip be taken into account. The radial force can be exerted either by a constant spring force or with the aid of a torque sensing device that adjusts the radial force in proportion to the torque loading. In the first case conditions similar to those when operating with a constant belt tension, section 14.4, are obtained, while in the latter case conditions corresponding to those in self-adjusting belt tensioning devices, section 14.7, are obtained. With torque sensing devices conditions can be chosen that give a safety margin against excessive slip. If the torque sensing device is placed on the driving shaft, then a somewhat higher radial force should be selected than if it is placed on the driven shaft.

15. Fatigue and belt life

15.1 Introduction

The life of V-belts will under normal conditions, i e when the belts are not rapidly destroyed by abnormal wear, excessive heating or extreme overloading, be determined by their fatigue properties. Fatigue is here defined as the gradual, degrading effect of repeated, small mechanical stresses and strains, often in combination with chemical attack from the surrounding atmosphere. The small stresses and strains are not large enough to cause rupture in a single cycle of loading but by being repeated they will gradually lead to failure. Failure may appear as a weakening of the reinforcement or wrapping fabric, eventually leading to rupture, as separations between reinforcement and rubber materials, as separations between various rubber components, as transverse cracks growing from the top or bottom side of the belt or as longitudinal cracks growing from the surface of the belt, cf figure 9-15. Even if fatigue failure in the adhesion between various components or in the rubber material is the weak link in the fatigue process calculations of fatigue life are usually based on tensile and flexural forces in the reinforcement as will be shown below. An increase in these stresses, however, will at the same time be accompanied by increases in stress level at the adhesion layer and in the rubber material.

Fatigue rate is strongly affected by the working temperature of the belt – thus fatigue rate will be doubled (life cut in half) by an increase of 10°C in belt operating temperature. Fatigue is a complicated process, which does not only depend on loading conditions, variations in loading conditions or the surrounding atmosphere but also on the existence of small imperfections in the product, which may be the starting point for fatigue cracks to form. In this chapter some of the most important factors that affect fatigue rate and life of V-belts will be discussed as well as calculation methods that under simplifying assumptions may be used to predict and quantify the effect of fatigue phenomena.

184

15.2 Fatigue curves

Life to failure will depend on stress or strain level, e g calculated according to section 13.6. The higher the stress or strain level, the shorter the life. The graphical representation of the relation between stress or strain level and life is termed the fatige curve, also called the Wöhler curve. The two quantities can be plotted either on logarithmic or linear scales, but it is usually preferred to plot life, e g expressed in hours or number of deformation cycles to failure, on a logarithmic scale and stress or strain level on either logarithmic or linear scales, figure 15-1. The advantage of using logarithmic plotting is that linear fatigue curves will be obtained over a large interval which facilitates extrapolation to very long working lives and makes mathematical treatment easier. With stress level σ on a linear scale and working life N on a logarithmic scale according to figure 15-1 B, the relation can be expressed as:

$$N = N_{ref} \cdot 10^{(\sigma_{ref} - \sigma)q} \qquad (15:1)$$

The equation contains a reference level of stress σ_{ref} and corresponding life N_{ref}.

With both stress level and working life on logarithmic scale according to figure 15-1 C the following equation will be obtained for the fatigue curve:

$$N = N_{ref} \cdot \left(\frac{\sigma_{ref}}{\sigma}\right)^q \qquad (15:2)$$

Typical of the fatigue curve is often that it levels out at a low stress or strain level, usually called the fatigue limit, below which no fatigue failure can be observed. The fatigue exponent q usually has a value of 4–8. This means that even small changes in stress or strain level may give rise to large changes in life. Typically an increase in stress level σ of 10% will cut belt life in half.

15.3 Effect of stress level

The total tensile stress in a belt section is calculated according to section 13.6 as the sum of the stresses resulting from pre-tensioning, power transmission, centrifugal forces and bending. As fatigue is a dynamic phenomenon it should be observed that stresses may have different effects, depending on whether

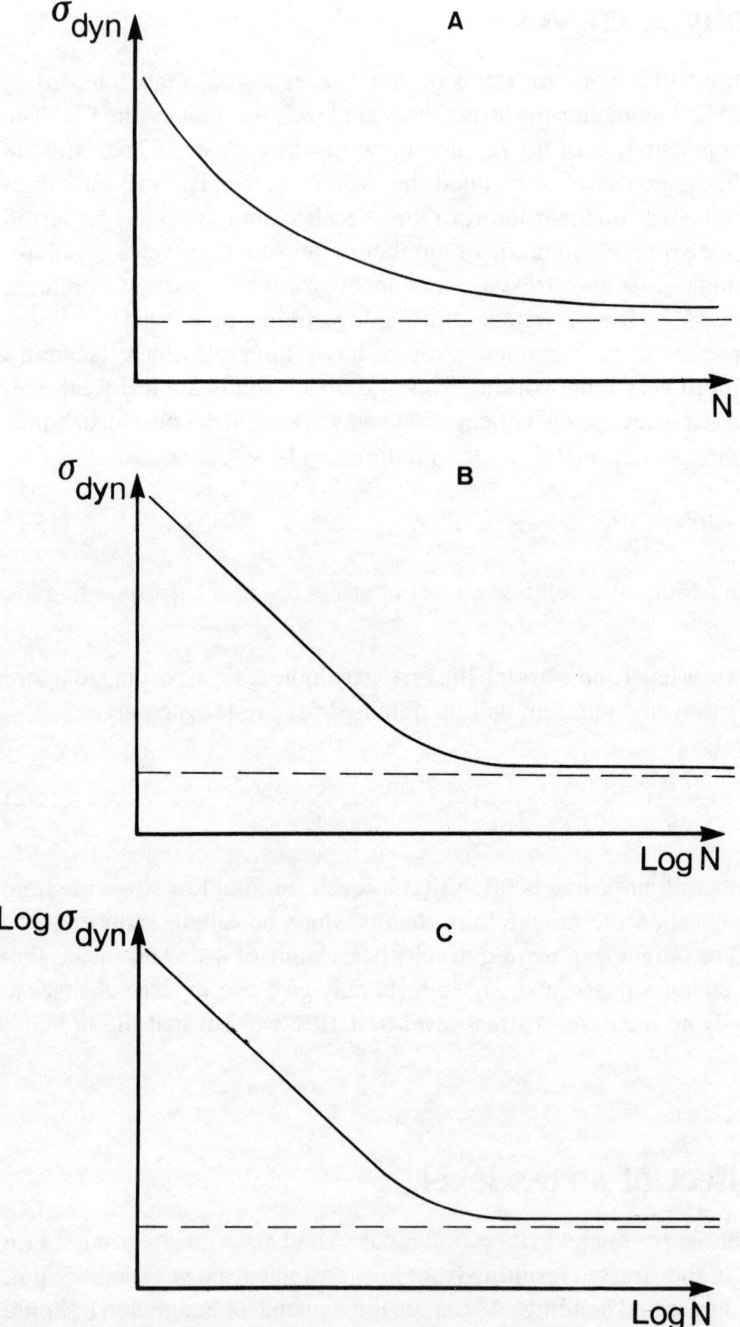

Figure 15-1. Fatigue curves showing the relation between number of deformation cycles to failure N and dynamic stress level σ_{dyn}. The dotted line indicates the fatigue limit
A. Linear scales   B. Semi-logarithmic diagram   C. Logarithmic scales

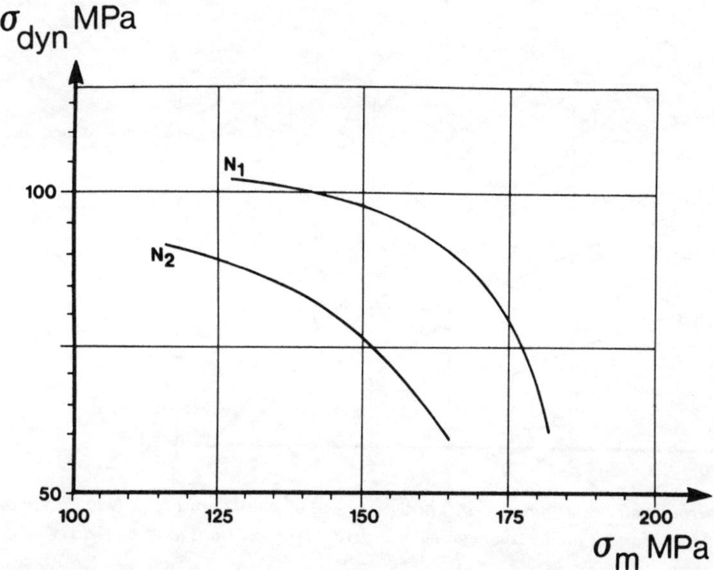

Figure 15-2. Experimentally determined V-belt life as a function of average tensile stress level (σ_m) and dynamic stress level (σ_{dyn}) in the reinforcement. The curve denoted by N_1 is for a belt life of 5×10^6 cycles, curve denoted by N_2 30×10^6 cycles (Oliver et al, 1976)

they can be described as dynamic or static. Stress level can then be divided into a static part – an average stress σ_m – and a dynamic part σ_{dyn} superimposed upon the static stress level. The dynamic stress σ_{dyn} is derived from the pulsation in belt tension force between tight and slack side of the belt and from bending stresses when the belt is bent over pulleys. Experimental investigations show that both static and dynamic stress levels are of importance for V-belt life, figure 15-2. An increase in static stress σ_m will thus only allow a lower dynamic stress σ_{dyn} for a given belt life and vice versa.

15.4 Effect of bending stresses

Simple methods for calculation of bending (flexural) stresses are discussed in section 13.4. The effect of flexural stresses on belt life will depend on the belt itself, mainly its stiffness (Young's modulus) and height, and also on pulley diameters, figure 15-3. Design calculations are usually based on the most severe case, i e the smallest pulley at which the highest flexural stresses are obtained. At large speed ratios lower, average fatigue stress levels will be

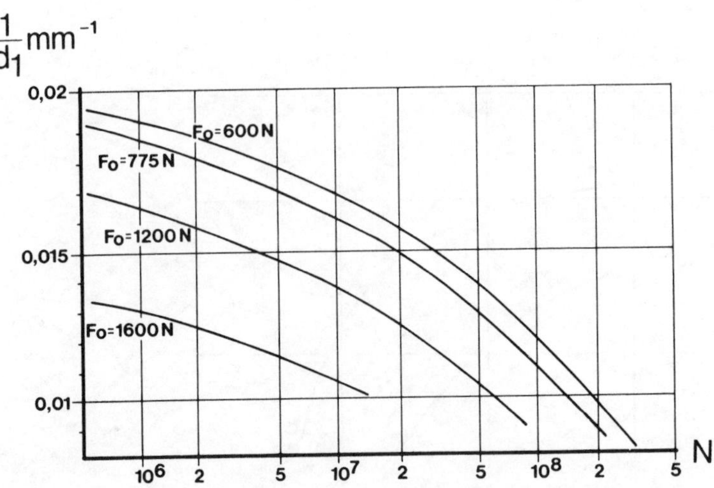

Figure 15-3. Experimentally determined V-belt life, expressed as number of cycles N to failure in relation to the inverse of small pulley diameter d_1 at different levels of shaft load F_0 (Oliver et al, 1976)

obtained than with two small pulleys of the same diameter. This will increase belt life, alternatively permit higher power ratings, cf section 16.7.

Another important factor is the number of flexures, which is inversely proportional to the length of the belt (or shaft centre line distance). The larger the shaft centre distance or belt length, the fewer the number of flexures needed to make a given number of shaft revolutions in a given time at constant belt speed. Not only the number of flexures will be reduced but also the number of tensile force changes between tight and slack side of the belt. Increase in belt length will therefore lead to longer belt life, alternatively increased power rating, cf section 16.9. If dynamic stress σ_{dyn} is the dominating factor in the fatigue process, then belt life will be directly proportional to belt length, cf section 15.7.

If the belt is bent in the same direction during its operation a certain relaxation will take place, which tends to even out the largest bending stresses. The belt should also be constructed to avoid concentrated bending stresses in any part of the belt section. If bending is reversed (double-bending), which is often the case when using idler pulleys, the counteracting bending stresses will increase fatigue rate and considerably shorten belt life. Transmission belts with low thickness (height), e g flat belts or V-ribbed belts, are less affected by double-bending than V-belts. One example of good and poor placement of an idler pulley, for long belt life, is shown in figure 15-4.

188

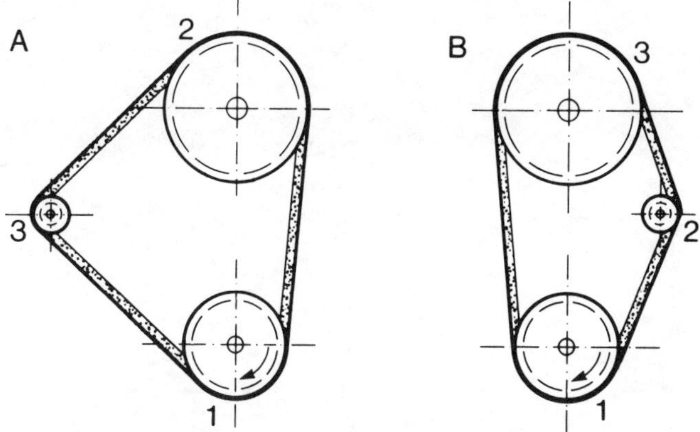

Figure 15-4. Examples of good (A) and poor (B) design for pulley arrangements in drives with more than two pulleys. Small pulleys should preferably be placed on the slack side to avoid the combination of large tensile and bending stresses. Pulleys should also be placed so that sufficient arc of contact angle is obtained whereby belt tension can be kept low. The driving pulley is denoted by 1

15.5 Variable load – cumulative damage theory

Typical of V-belts in service is that they are exposed to large load variations. This variation is partly caused by the varying stresses introduced when belts are bent over pulleys or when tension force level fluctuates between tight and slack side of the belt and partly from conditions resulting from variations in power requirement from the driven unit, e g during start-up, overloads, etc. For accurate calculations and estimations of belt life the effect of these varying stress and strain levels must be summarized, which for V-belts, like for many other dynamically stressed components, can be made by applying the cumulative damage theory. The conditions of loading are then described by a distribution curve, figure 15-5, in which the stress level, σ_i, is plotted against the number of cycles of deformation, N_i, at each stress level. At each level of loading the corresponding fatigue life N_a will then be reduced according to equation 15:2, resulting in:

$$\frac{N_i}{N_a} = \frac{N_i}{N_{ref}} \frac{\sigma_i{}^q}{\sigma_{ref}{}^q} \quad\dots\dots\dots\dots\dots\dots\dots\dots\dots\dots\dots\dots\dots (15:3)$$

According to the linear theory of cumulative damage the sum of these fractions is equal to 1, i e:

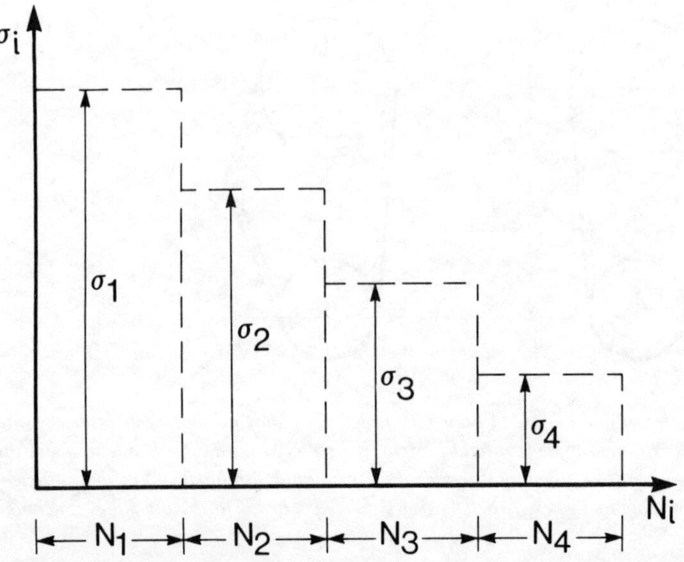

Figure 15-5. Example of stress distribution in time for a V-belt in operation

$$\Sigma(N_i\,\sigma_i{}^q)=N_{ref}\,\sigma_{ref}{}^q \quad\text{...} (15:4)$$

The fatigue life N corresponds to an average equivalent stress σ_e according to:

$$\sigma_e=\sqrt[q]{\dfrac{N_1}{N}\,\sigma_1{}^q+\dfrac{N_2}{N}\,\sigma_2{}^q+\,.\,.\,.} \quad\text{...} (15:5)$$

15.6 Belt life variations

Typical of belt life, as for the life of e g light bulbs or human beings, is that it never assumes a fixed value for a group of individuals, but shows a usually rather wide spread. Typically 10% of the belts will fall below half the medium life of the population while about 15% will surpass 1,5 times the medium life. It is thus not unusual that individuals within a lot of modern V-belts may show variations in belt life of up to a factor of 3. For earlier V-belts, manufactured from materials with large variations in properties and with less controlled manufacturing techniques, lifetime variations of up to a factor of 10 were not

190

Accumulated percentage of failure	Belt life, h
11	228
22	331
33	337
44	465
56	503
67	584
78	585
89	647

Figure 15-6. Results from fatigue life testing of a number of V-belts from the same lot in a dead-weight machine. The percentage failure is given for the number of belts plus one (data from Trelleborg AB)

uncommon. This is mostly explained by reference to micro-defects that appear at random in the material and may be the starting point for fatigue cracks, or to variations in service conditions, which are never constant. Belt life is then usually distributed according to the so-called Weibull distribution. This differs from the normal (Gaussian) distribution in that it may have a certain skewness to higher or lower values. In figure 15-6 the results from fatigue testing of a number of V-belts from a certain lot under identical conditions are given. To get maximum information out of the data they are plotted on double-logarithmic scales on a graph paper with scales according to the Weibull distribution, figure 15-7 (printed diagrams are for sale). If the results are to follow the Weibull distribution they may be fitted to a straight line with good accuracy. From the curve a typical belt life can be calculated – a type of average life – and also a figure giving the spread of the distribution. The probability that belt life may fall below a certain level may also be read off from the graph. Statistical treatment of data in this way can also shorten testing time considerably. On a set of V-belts to be tested it is not necessary to wait until the test with the belt with the longest life has been completed. It is sufficient to run the test until most belts have failed and then plot the Weibull graph with acceptable accuracy.

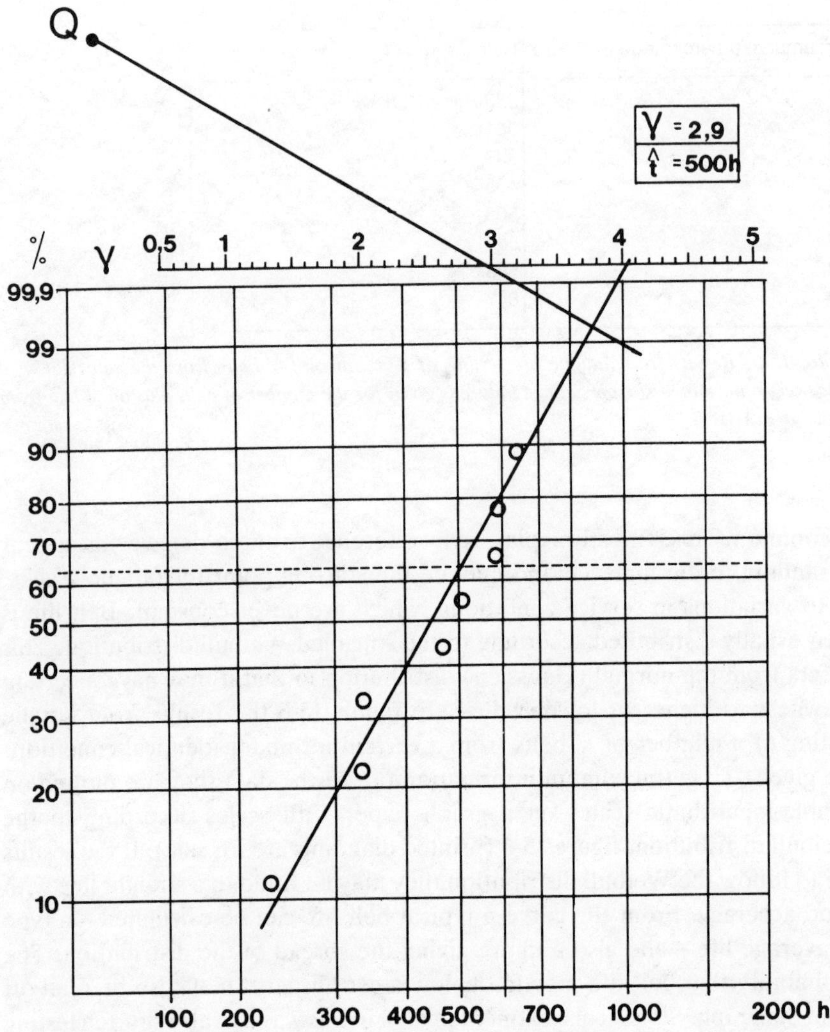

Figure 15-7. *Weibull plot of fatigue data from figure 15-6. Percentage failure is plotted against belt life (h). A characteristic belt life (a type of median value) t̂ is calculated for 63% failure. If a straight line is drawn normal to the curve from point Q a value γ can be read off a scale which is a measure of the spread in values and the shape of the distribution curve*

15.7 Semi-empirical equations for estimating belt life

Fatigue processes in V-belts are rather complicated and difficult to treat quantitatively and mathematically. Design calculations are usually based on theoretical considerations and the equations thus obtained modified with empirical coefficients, the value of which will depend on type of belt, belt dimensions and other things. To estimate these coefficients, laboratory fatigue tests are run according to section 8.3. The results obtained are then treated statistically with regression analysis, often performed with the aid of computers, whereby the best estimates for all coefficients are obtained. As the resulting equations are derived by combining theoretical considerations and experimental results they may be termed semi-empirical.

The theoretical calculations start from the assumption of a maximum total stress allowed in the belt or reinforcement, e g calculated with equation 13:8. This allowed maximum total stress will include the belt tension required for pre-tensioning and power transmission, flexural stresses when passing over the pulleys as well as stresses caused by centrifugal forces. The maximum total stress thus calculated must then be reduced with regard to the number of deformation cycles, i e desired belt life, e g by application of equations 15:1 or 15:2. If total stress will exist at different levels, e g when passing over pulleys of different diameters, its effect must be summarized by applying the cumulative damage theory, e g by using equation 15:4. Examples of different methods of calculation, mainly developed within V-belt manufacturing companies, are presented below.

One method of calculation proceeds from the stress level at the most heavily loaded cord thread, usually the one closest to the upper edge of the belt. The shear forces needed to transform the power from pulley groove sides to the reinforcement have here their maximum, cf figure 12-7. The stresses cannot be relieved by sagging of the cords, which is possible for the central cords of the V-belt, cf figure 3-4. The difference in stress level between cords at the edges and the average cord tension will increase with increase in ratio between the stiffness of the cord material and rubber. This is considered during calculations by the inclusion of a coefficient, the value of which will depend on the stiffness ratio. Stresses in the rubber will follow those in the cords, meaning that the theory is not limited to the fatigue of the reinforcement but will also include fatigue failure in the rubber. The method is applicable for drives with a large number of pulleys and also for those with a large number of loading levels. The fatigue life N_{ij} at the i:th stress level and the j:th pulley is calculated according to:

$$N_{ij}=k_1 \, (k_0-\sigma_{dyn \, ij})^2 \, (k_m-\sigma_{m \, ij})^2 \, L^{1.75} \, v^k \dots\dots\dots\dots\dots\dots\dots (15\!:\!6)$$

Here $\sigma_{m \, ij}$ denotes the average (static) stress level in the cord and $\sigma_{dyn \, ij}$ the dynamic stress component, resulting from bending stresses and fluctuations in belt tension force between slack and tight side. L denotes belt length and v belt speed. Belt life is then calculated by summing up the fatigue effects at the various pulleys by applying the cumulative damage theory.

Another calculation method works with fatigue rate (FR), instead of belt life. Fatigue rate FR_j at the j:th pulley is then calculated according to:

$$\log FR_j=k_1 \, (F_{e1j}+\frac{k_2}{d_j}+k_3 \, v^2)+k_4 \log v+k_5 \dots\dots\dots\dots (15\!:\!7)$$

Decisive factors for the fatigue rate are thus effective tight side tension F_{e1j}, pulley diameter d_j and belt speed v. Belt life N is then calculated by summing the fatigue rates at various stress levels:

$$N=k\frac{L}{\Sigma FR_j} \dots\dots\dots\dots\dots\dots\dots\dots\dots\dots\dots\dots\dots\dots (15\!:\!8)$$

With this method of calculation belt life is directly proportional to belt length L.

In the equations above k with various indexes denotes coefficients that have to be determined experimentally for each belt type. Calculations can only be performed if these coefficients are known. They can for instance be found in belt manufacturers' catalogues. Here more detailed descriptions of the calculations are also given. Many of the calculations are available as computer programs.

15.8 Required belt life

Requirements on belt life vary between different application areas. For industrial drives, which usually operate during a large part of the day and night, a belt life of at least 25 000 h is usually required, corresponding to 3–5 years of use for the equipment. For intermittent and seasonal applications, e g in agriculture and household, a considerably shorter belt life can be accepted while retaining the requirement of undisturbed use of the equipment

Application	Typical belt life requirement, h
Industrial	
Equipment with long operating hours or continuous operation, e g fans, pumps or conveyors	12 000–25 000
Hand tools, office equipment	7 500–20 000
Equipment with intermittent or occasional operation	6 000–12 000
Agricultural	
Stationary equipment with long operating hours or continuous operation, e g conveyors, pumps or fans	6 000–12 000
Stationary installations with intermittent operation	2 000– 6 000
Mobile equipment, e g harvesters, sowing-machines, manure spreaders or hay balers	500– 1 000
Automotive	
Passenger cars, vans	1 000– 3 000
Lorries, trucks	5 000–10 000
Buses, tractors, road construction machines	5 000–10 000
Home appliances	
Heating, ventilation, air-conditioning	5 000–10 000
Washing machines, tumbler dryers, dish-washers	1 500– 2 000
Sewing machines, lawn-mowers, hand tools	200– 1 000

Figure 15-8. Normal belt life requirements in some applications

for 3–5 years. Figure 15-8 summarizes typical requirements on belt life for some applications. As evident from the discussion above there are no technical problems involved in designing drives for considerably longer belt lives. This over-dimensioning will, however, increase the cost for the installation and also increase its weight and space requirements. The belt lives suggested in figure 15-8 are therefore typical compromises with regard to the simultaneous requirements of compact design and low price.

16. Power transmission

16.1 Introduction

In this chapter the various factors that affect the power transmission capability of a belt drive will be discussed. These involve the dimensioning and construction of the V-belt as well as factors in the drive, primarily diameters and rotation speeds of pulleys and shaft centre line distance. Special conditions in service will also be considered, by utilizing the so-called service factor.

16.2 Basic power relations

The power P that can be transmitted in a belt drive is the product of angular velocity ω and torque M:

$$P = \omega \cdot M \hspace{3cm} (16:1)$$

This equation expresses an old experience, i e that a certain power can be transmitted with either high rotation speed in combination with a low torque or a high torque in combination with a low rotation speed. Power P can also be written as the product of the difference between tight side and slack side tension force F and belt speed v:

$$P = F \cdot v = (F_1 - F_2) v \hspace{2cm} (16:2)$$

This equation shows in the same way that a given power can either be transmitted by combining a high belt tension force difference with a low belt speed or a high belt speed with a low belt tension force difference. If pulley diameter d is introduced, equations 16:1 and 16:2 can be combined into:

$$P = \omega \cdot \frac{d}{2} \cdot F \hspace{3cm} (16:3)$$

As centrifugal forces will give the same tension force increment in both slack and tight side of the belt they will not affect the belt tension force difference, which can then be written (F_e denotes the effective belt tension):

$$F = F_{e1} - F_{e2} = F_1 - F_2 \quad\text{...} \quad (16\colon 4)$$

16.3 Dimensioning and number of belts

The decisive factor for the power rating of the V-belt is the capability of the belt to take up high tension forces. The more reinforcement it contains and the better the properties of the rubber and reinforcement material are, the higher its power rating will be. Introducing more reinforcement requires more space and a larger belt section area. The largest tension force F_1 is found in the tight side of the belt. Taking dynamic and fatigue conditions into account a large safety factor has to be used when dimensioning relative to the static tensile strength of the belt, for instance a fourfold safety factor. Earlier V-belts manufactured under less stringent conditions and with larger variations in material properties, required much higher safety factors, e g a tenfold safety factor. This gave heavier and more expensive constructions, and lower mechanical efficiency. Figure 16-1 shows how dimensioning has been carried out in some practical cases. The static tensile strength of a belt amounts roughly to 75–95% of the sum of the static tensile strengths of the individual cords. Correct placement of the cords will give an even stress distribution and more efficient utilization of cord strength. The values obtained in tensile testing as in figure 16-1 will strongly depend on the testing procedure used. Fabric reinforcement will lead to less efficient utilization, as the cords or

Profile	Number of cords per belt (A)	Tensile breaking load per cord (B), N	Theoretical tensile breaking load of belt (C)=(A)·(B), N	Measured tensile breaking load of belt (D), N	Efficiency of cord strength utilization $\frac{(D)}{(C)}$	Max permitted belt tension (E), N	Saftey factor $\frac{(D)}{(E)}$
SPZX	7	450	3150	2520	0,80	500	5,0
SPAZ	7	730	5110	4500	0,88	850	5,3
SPBX	7	1080	7560	6100	0,81	1500	4,1

Figure 16-1. Examples of reinforcement utilization in the design of V-belts (data from Trelleborg AB)

197

threads in the warp direction (transverse to the length direction of the belt) cannot take up tension forces in the belt.

The method used in transmitting large power ratings is usually not to use one single belt of heavy section but to use several belts of smaller section running in parallel on pulleys with several grooves, so-called multiple belt drives. The power rating of a multiple belt drive will depend on how well the load can be distributed evenly among the individual belts. With modern manufacturing techniques, belt length tolerances are so small that it is usually assumed that a multiple belt drive can transmit a power equivalent to the sum of the power ratings of the individual belts, cf section 18.9.5. Some methods of calculation still apply a certain reduction for multiple belt drives. Thus it may be assumed that for a multiple belt drive with 2–3 belts only 95% of the sum of the individual power ratings can be utilized, with 4–6 belts 90% and with more than 6 belts 83%.

16.4 Belt tension forces

The power transmitted will according to equation 16:2 depend on the difference between belt tension forces in tight and slack side respectively. The ratio between effective belt tension forces in the tight and slack sides can be expressed by Eytelwein's equation, equation 12:16. For belt drives operating with power transmission by friction a certain slack side belt tension will thus be required, while belts that transmit power by mechanical interlocking, e g synchronous belts, at least theoretically need no belt tension in the slack side. By combining the two equations we obtain:

$$P = (F_{e1} - F_{e2})\, v = F_{e2} \left(1 - e^{\mu_s \theta}\right) v \qquad \text{(16:5)}$$

This equation demonstrates the importance of slack side effective belt tension F_{e2} for adequate power transmission.

The relation between effective belt tension forces and the difference between them can be expressed by different ratios. The simplest one is the ratio $F_{e1}{:}F_{e2}$, which for V-belt drives should preferably not exceed 5:1. Another ratio is the effective belt tension ratio λ, defined according to:

$$\lambda = \frac{F_{e1} - F_{e2}}{F_{e1}} = 1 - \frac{F_{e2}}{F_{e1}} \qquad \text{(16:6)}$$

Furthermore a ratio termed the coefficient of traction λ_b can be used, defined according to:

$$\lambda_b = \frac{F_{e1} - F_{e2}}{F_{e1} + F_{e2}} \quad\dotfill (16\!:\!7)$$

The ratios λ and λ_b will vary between 0 and 1. At value 0 there will be no difference between tight and slack side belt tension force and thus no power transmitted. Value 1 corresponds to the case when slack side belt tension force is infinitely small compared to tight side belt tension force, and can in service only be reached with belts using mechanical interlocking, e g synchronous belts. For $F_{e1}\!:\!F_{e2}=5\!:\!1$ λ will be 0,8 and λ_b 0,67. In a belt drive pre-tensioned at constant elongation, the shaft loading $F_0=F_{e1}+F_{e2}$ is constant at varying torque loading M. As $F_{e1}-F_{e2}=\dfrac{2M}{d}$ the coefficient of traction λ_b will be directly proportional to the torque loading. The coefficient of traction will therefore be a measure of the effective utilization of the belt drive in relation to the pre-tension applied. A well designed belt drive will be utilized to 67%, as shown above. A high value of ratio λ or λ_b is desirable in a belt drive. This will mean a low average belt tension force, leading to longer belt life and reduced loads on shafts and bearings.

16.5 Belt angle and friction

If equations 16:6 and 16:7 are combined with Eytelwein's equation 12:16, then the following relations will be obtained:

$$\lambda = 1 - e^{-\mu_s\theta} \quad\dotfill (16\!:\!8)$$

$$\lambda_b = \frac{e^{\mu_s\theta} - 1}{e^{\mu_s\theta} + 1} \quad\dotfill (16\!:\!9)$$

These equations demonstrate the strong dependence of ratios λ and λ_b on the product of the apparent coefficient of friction μ_s and arc of contact angle θ. This relation is illustrated graphically for $\theta=180°$ in figure 16-2. According to equation 12:3 the apparent coefficient of friction μ_s is defined as the ratio of the coefficient of friction μ and sinus for half the belt angle α. This means

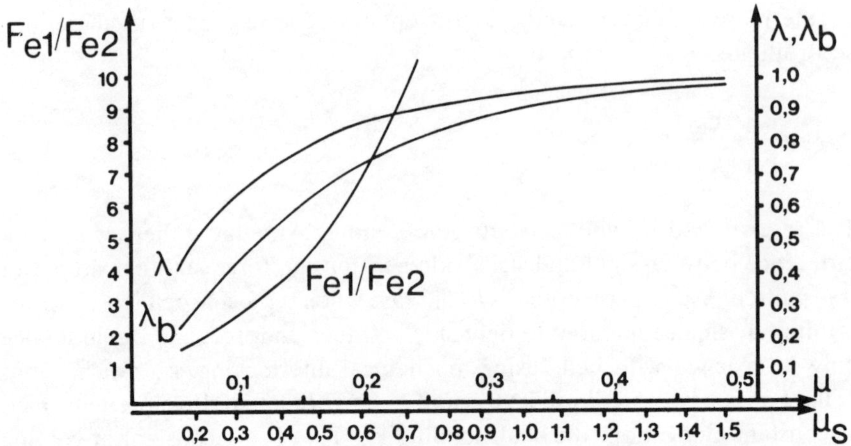

Figure 16-2. Ratios F_{e1}:F_{e2}, λ and λ_b as function of coefficient of friction μ and apparent coefficient of friction μ_s at a belt angle=38° and arc of contact angle=180°

that V-belts can operate at much higher values of λ or λ_b than for instance flat transmission belts, which are usually designed for F_{e1}:F_{e2}=2:1.

If power transmission in a belt drive is increased, then a point will sooner or later be reached where excessive belt slip occurs. The simplest way to counteract excessive slip in a given belt drive is to increase pre-tension of the belt, whereby slack side belt tension force will be increased, leading to a larger belt tension force difference. Other methods, normally limited to completely new installations, include the use of belts with materials with higher coefficients of friction or belts with reduced belt angles. The widely used standardization of V-belts and pulleys will, however, restrict the selection of other belt angles than the standardized ones. From figure 16-2 can be seen that an increase in the coefficient of friction to above 0,5 or in the apparent coefficient of friction to above 1,5 will lead to only marginal increases in power transmission capability at an arc of contact angle of 180°.

16.6 Belt speed and rotation speeds of pulleys

To study the influence of belt speed or rotation speed, expressed as angular velocity ω, when pre-tensioning at constant elongation, equations 16:1 and 16:2 are first considered. These equations demonstrate that the power transmitted is directly proportional to speed. The validity is, however, not unlimited. In section 13.3 is demonstrated how the increase in centrifugal

200

forces with increase in speed will limit available effective belt tension forces. By applying the simplifying assumptions, leading to eqution 13:8, the effective tension stress σ_e in the belt can be expressed as:

$$\sigma_e = \sigma_t - \sigma_b - \sigma_c = \sigma_t - \frac{ET}{d} - \varrho\, v^2 \quad\text{...} \quad (16{:}10)$$

Here σ_t denotes the total stress that can be permitted in the belt with regard to its ultimate strength, σ_b the tension stress caused by bending the belt over pulleys and σ_c the tension stress caused by centrifugal forces. The belt is assumed to be made from a homogeneous material with Young's modulus E and density ϱ. The height of belt section is denoted T, and the diameter of the smallest pulley is d=2r. If belt section area is A the power transmitted P by combination of equations 13:1, 16:2, 16:6 and 16:10 can be written as:

$$P = \lambda A \sigma_e v = \lambda A \left(\sigma_t - \frac{ET}{2r} - \varrho v^2\right)v = \lambda A \left(\sigma_t - \frac{ET}{2r} - \varrho\omega^2 r^2\right)\omega r \quad\text{.........}\quad (16{:}11)$$

This equation shows that the power transmitted initially rises in direct proportion to belt speed v or angular velocity ω. As the influence of centrifugal forces becomes increasingly important the curve flattens out, reaches a maximum and then turns downwards. The critical factor is the small pulley with the angular velocity ω_1. To obtain the maximum value of the power transmitted, P_{max} and corresponding angular velocity $\omega_{1\,opt}$, derive P in the equation with respect to ω and assume a sufficiently large and constant value of d:

$$\omega_{1\,opt} = \sqrt{\frac{\sigma_t - \dfrac{ET}{2r_1}}{3\varrho r_1^2}} \quad\text{..}\quad (16{:}12)$$

$$P_{max} = 2\lambda A\varrho \left(\frac{\sigma_t - \dfrac{ET}{2r_1}}{3\varrho}\right)^{1,5} \quad\text{.....................................}\quad (16{:}13)$$

The relation is demonstrated graphically in figure 16-3. For more accurate determinations practical tests have to be run. Examples are given in figure 16-4.

As the power transmitted within a wide range increases roughly in proportion to speed, a speed not too different from the optimal one is usually strived for in belt drive design. Figure 4-7 gives guide-lines on the maximum belt speeds

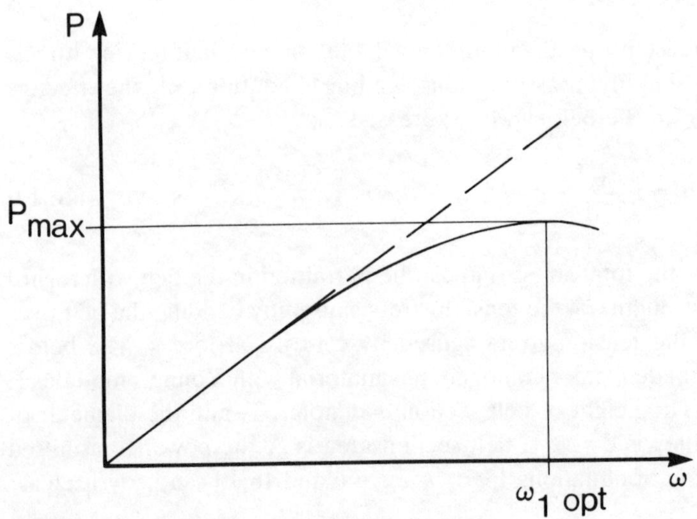

Figure 16-3. The power rating P that can be transmitted as a function of angular velocity ω at constant diameter d_1 of small pulley

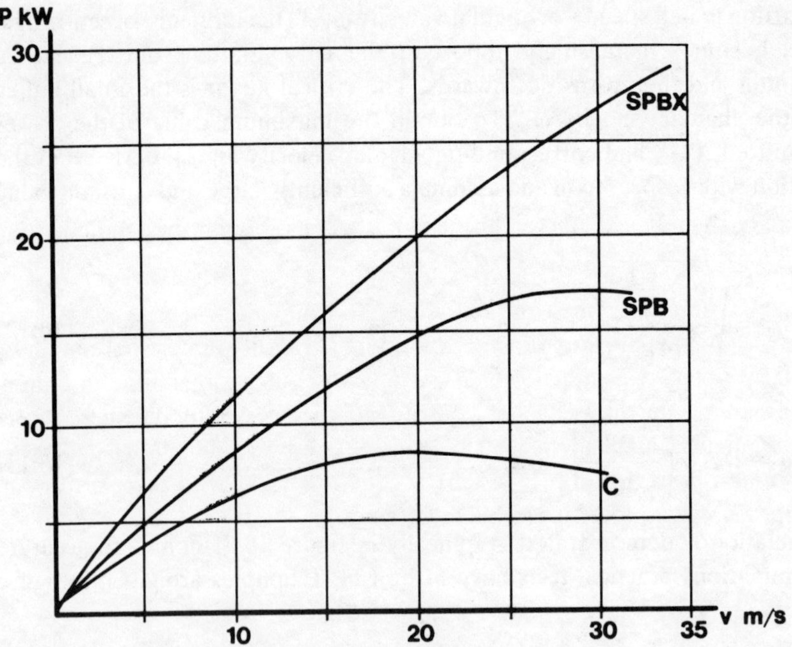

Figure 16-4. Experimentally determined data for power rating P for classical V-belt of C profile, narrow V-belt of SPB profile and narrow V-belt with cogged underside of SPBX profile as a function of belt speed v. The diameter of the small pulley is 180 mm (W Erickson, 1979)

recommended. For more accurate recommendations V-belt manufacturers' catalogues should be consulted. At low rpm's, let us say at belt speeds below 5 m/s, belt drives are generally not competitive with other types of power transmission.

16.7 Pulley diameter

In a belt drive small pulleys will give lighter and cheaper constructions requiring less space. At a given shaft rotation speed a reduction in pulley diameter will lead to a decreased belt speed. This may be a drawback at low rotation speed, as power transmission will be reduced, but at high rotation speed it may be an advantage as the risk of exceeding the maximum permitted belt speed will be eliminated. As belt speed at constant angular velocity increases in direct proportion to pulley diameter the corresponding increase in centrifugal force will put an upper limit on pulley diameter. On the other hand decrease in pulley diameter will lead to increased bending stresses, whereby the available effective tension stress will be reduced, cf equation 16:11. Below a certain pulley diameter bending stresses will become so large that belt life may be drastically reduced. For most belts and belt drives values for minimum recommended pulley diameter can be found, see tables in chapter 3. Low height T of the belt section and cogging of its underside will increase the flexibility of the belt and its ability to run over small diameter pulleys.

The effect of pulley diameter on power transmission can be studied from equation 16:11.

At pulley diameters above the recommended minimum (corresponding to radius $r_{1\ min}$) a proportional increase in power transmission is first obtained (angular velocity constant), which then levels out at an optimum value, determined by the influence of centrifugal forces. Additional increase in pulley diameter will then only lead to a decline in power rating. For calculation of maximum power rating P_{max}, derive equation 16:11 with respect to r (at constant ω), which after rearrangement gives:

$$r_{1\ opt} = \sqrt{\frac{\sigma_t}{3\varrho\ \omega_1^{\,2}}} \quad\dotfill\quad (16:14)$$

$$P_{max} = \lambda A\left[2\varrho\left(\frac{\sigma_t}{3\varrho}\right)^{1,5} - \frac{ET\omega}{2}\right] \quad\dotfill\quad (16:15)$$

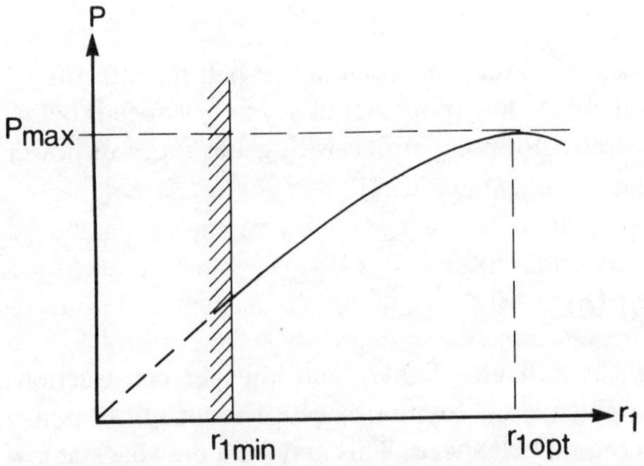

Figure 16-5. Power that can be transmitted P as a funtion of small pulley radius r_1 at constant angular velocity

A schematic, graphical representation of the equation is given in figure 16-5. Published experimental studies are seldom extended above belt speeds, where the influence of centrifugal forces becomes predominant, why only an increase in power rating with increased pulley diameter is given, cf example in figure 16-6.

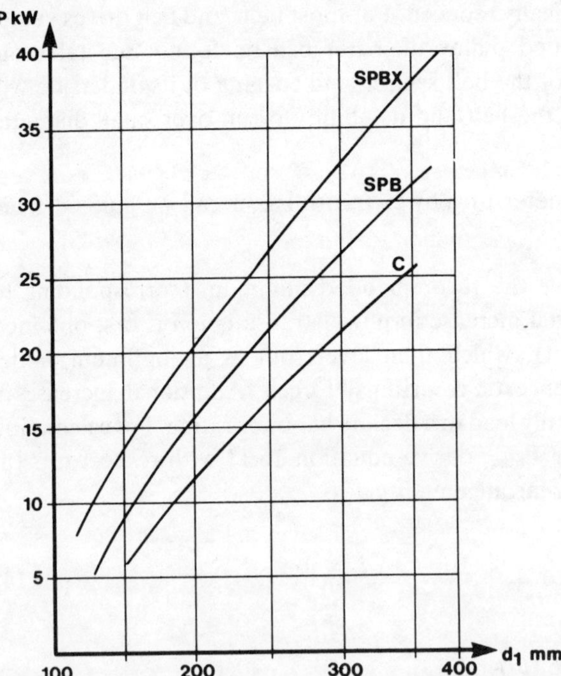

Figure 16-6. Experimentally determined data of power rating P for classical V-belt with C profile, narrow V-belt with SPB profile and narrow V-belt with cogged underside with SPBX profile as a function of small pulley diameter d_1. The experiments were carried out at a fixed angular velocity 1750 rpm of the small pulley (W Erickson, 1979)

16.8 Speed ratio and arc of contact angle

In section 15.4 it is mentioned that large speed ratios in a belt drive will give low bending forces at the large pulley, leading to increased belt life or alternatively higher power rating. This is compensated for in belt design calculations by introducing an add-on power $\triangle P_{b1}$, the value of which will increase with speed ratio, cf section 18.9.2. With increase in speed ratio the arc of contact angle at the small pulley will decrease, which, however, acts in the opposite direction. A correction factor smaller than 1 has to be introduced in belt design calculations, cf section 18.9.4. Increasing the arc of contact angle with the aid of idler pulleys on the outside of the belt may not necessarily have a positive effect, as the double-bending from the idlers will increase fatigue stresses, leading to reduced belt life alternatively a lower power rating, cf section 19.3.

16.9 Belt length

The positive effect of increased belt length on belt life is discussed in section 15.7. The reduction in the number of flexures and number of stress changes between the slack and tight side of the belt, as well as of working temperature, normally leads to increased belt life. The increase is at least in proportion to belt length, but normally to belt length to a higher exponent, cf equations 15:6 and 15:8. With a given belt life alternatively a higher power can be transmitted. In design calculations for belt drives an add-on power for belt length can be introduced, cf section 18.9.3, taking belt length and type of belt into account. Alternatively a power correction factor can be used, the magnitude of which depends on the ratio between actual length and the base length, for which the basic power rating is calculated.

16.10 Service factor

The basic power rating for a belt is valid for a drive, not exposed to excessive variations in loading or other external disturbances. In service, load can vary considerably, e g during start-up of the belt drive or with load variations from driving or driven units. Electric motors usually give a constant load, with exception for start-ups, while combustion engines, e g diesel engines, usually operate with pulsating torque. In the same way a driven unit, like a blower, may operate with a constant torque, with the exception for start-ups, while other driven units, like crushers, compressors, etc, show large torque variations. Other factors that increase the severity rating in a drive are long daily working hours, which lead to increased belt operating temperatures, and chemical or other environmental effects, etc. The increased severity rating due to all these factors, which will reduce the expected belt life, is taken into account by introducing the so-called service factor in design calculations, cf section 18.4.

16.11 Quantitative calculation of power rating

As with calculation of belt life, cf section 15.7, equations may be developed for calculation of power ratings, which partly are based on theoretical considerations and then modified with coefficients from experimental investigations. The equations can then be characterized as semi-empirical and contain a number of coefficients, the values of which have to be determined experimentally for each belt type, each belt profile section and desired belt life. In the calculation method suggested in ISO 5292 the total power rating P of the drive is given, as the sum of the basic power rating P_b, the add-on power for speed ratio $\triangle P_{b1}$ and the add-on power for belt length $\triangle P_{b2}$ according to:

$$P=k \; (P_b+ \triangle P_{b1}+ \triangle P_{b2}) \quad\dotfill (16{:}16)$$

The coefficient k, which takes care of variations in arc of contact angle, is calculated according to:

$$k=1{,}25 \; (1-5^{-\theta/\pi}) \quad\dotfill (16{:}17)$$

The basic power rating P_b is calculated according to:

$$P_b=d_p \; \omega \left[k_1-k_2 \frac{1}{d_p}-k_3 \; (d_p \; \omega)^2-k_4 \; \log \; (d_p \; \omega) \right] \dotfill (16{:}18)$$

The resemblance between this equation and equation 16:11 should be noted. The first term in bracket refers to the maximum allowable stress in the belt determined by its ultimate strength, the second one gives the reduction due to flexural stresses, the third one the reduction due to centrifugal forces, and

206

the fourth one the reduction due to fatigue stresses related to rotational frequency and belt speed.

The add-on power for speed ratio $\triangle P_{b1}$ is calculated according to:

$$\triangle P_{b1} = k_4 \omega \; d_p \; \log \frac{2}{1 + 10^{\frac{k_2}{k_4} \frac{1}{d_p} (\frac{1}{S} - 1)}} \qquad\qquad (16{:}19)$$

Here S is the largest value of either c_p or $\frac{1}{c_p}$, c_p being the speed ratio.

The add-on power for belt length $\triangle P_{b2}$ is calculated according to:

$$\triangle P_{b2} = d_p \; \omega \; k_4 \; \log \frac{L}{L_0} \qquad\qquad (16{:}20)$$

In this equation actual belt length L is related to a base length L_0. In design calculations the add-on power $\triangle P_{b2}$ is often replaced by a correction factor, cf equation 18:5 and section 18.9.3.

In the equations above pitch diameter d_p and angular rotation speed ω refer to the smallest pulley, which is the most critical one in this connection. Coefficients k_1, k_2, k_3, and k_4 are determined experimentally for each V-belt dimension. The coefficients are also related to a given belt life, usually 25 000 h.

The use of the equations above assumes that the values of all coefficients are known. More convenient is therefore to use the tabulated data for basic power rating P_b, usually presented as a function of rotation speed and diameter of the small pulley, which are to be found in many standards, cf references in the table in Appendix C. As improved V-belts are continuously being developed and marketed, the power ratings presented in standards often tend to be out of date. It may therefore be advantageous to use corresponding ratings taken from the V-belt manufacturers' catalogues, which refer to the most modern, improved types, cf example in Appendix G. The add-on effects for speed ratio $\triangle P_{b1}$ and belt length $\triangle P_{b2}$ (alternatively often expressed as correction factors) can also be obtained either from standards or from V-belt manufacturers' catalogues, cf examples presented in figures 18-7 and 18-8. The increased use of computers or microcomputers for belt design calculations make, however, equations of the type given above easier to use than tabulated figures of power rating.

17. Mechanical efficiency

17.1 Introduction

In chapter 16 belt drives are discussed as if they were completely free from losses, either in speed, torque or power. The normally small losses, that cannot be avoided, will be discussed in this chapter and quantified as the corresponding efficiency. Efficiency is then defined as the ratio between the real figure obtained in relation to the expected one, either expressed as a percentage or a fraction. Expressed as a percentage efficiency will vary within the interval 0–100% and expressed as a ratio within the interval 0–1. Despite the small energy losses involved in V-belt drives efficiency has become of increasing importance due to the increased scarcity and cost of energy.

17.2 Efficiency in speed transmission

In a belt drive without losses speed ratio c_p can be calculated as the ratio between pulley diameters d, torques M, angular velocities ω and speeds of rotation n of the two pulleys according to:

$$c_p = \frac{d_2}{d_1} = \frac{M_2}{M_1} = \frac{\omega_1}{\omega_2} = \frac{n_1}{n_2} \quad \text{..} \quad (17:1)$$

The validity of equation 17:1 is based on the assumption that a constant elongation over the whole length of the belt is at hand in the drive. Transmission belts are, however, manufactured from elastically deformable materials and will change their elongation under service conditions in proportion to the tension load applied. The difference between tight side tension force F_1 and slack side tension force F_2 will lead to a difference in elongation ε_d and a corresponding slip s, which for a belt of Young's modulus E and section area A can be calculated according to equation 17:2. Slip s is approximately equal to the elongation difference ε_d under the assumption that shear deformation in the rubber material between reinforcement and pulley groove sides as well as radial movement are neglected.

$$\varepsilon_d = \frac{F_1 - F_2}{E \quad A} \approx s \qquad \dots\dots\dots\dots\dots\dots\dots\dots\dots\dots\dots\dots\dots\dots\dots\dots\dots\dots \quad (17{:}2)$$

From equation 17:2 it is observed that the elongation difference is directly proportional to the belt tension force difference, which in turn according to equation 16:5 increases in proportion to the power transmitted. When the belt leaves the driven pulley on the tight side belt tension force will increase from F_2 to F_1, causing an increase in elongation. As is discussed in section 12.2 and illustrated in figure 12-3 this increase in elongation will give rise to a tangential slip in the pulley groove in the so-called slip zone. The slip zone θ_g is a part of the total arc of contact angle θ. In the same way the slack side of the belt, when leaving the driving pulley, will change its belt tension force from F_1 to F_2, leading to a corresponding reduction in elongation and slip in the slip zone of the driving pulley. The slip zones will increase in size with increases in power transmitted. When slip zones embrace the total arc of contact, friction force has been utilized fully. Attempts to additionally increase the power transmitted will lead to excessive slip, which we may call "sliding slip" or total slip, and which is superimposed upon the slip caused by difference in elongation, which we may call "tension slip". For belts with normal extensibility (Young's modulus) tension slip is usually of the order of 1% (experimental data usually within 0,5–2%). Take as an example a narrow V-belt with SPAX profile with a modulus of elasticity E=550 MPa and cross section area A=90 mm^2 and designed to transmit a force difference $F_1 - F_2$ of maximum 500 N. With these figures inserted in equation 17:2 s=1,0% is obtained. Sliding slip can increase catastrophically with increase in power requirement and may lead to rapid wear, excessive heating and drastic shortening of belt life. In figure 17-1 some experimental results are shown, giving the slip for wrapped and cut V-belts respectively as a function of torque M, small pulley diameter d_1, shaft load $F_1 + F_2$ and arc of contact angle θ. As expected slip increases with increases in torque, and decreases with increases in arc of contact angle. Increases in shaft load (pre-tension of belt) and small pulley diameter will also lead to decreased slip. In this investigation wrapped V-belts gave somewhat higher slip than cut V-belts.

If the belt is tested over a large torque interval, three regions can be distinguished: in the first slip will increase in direct proportion to the torque transmitted, in the second, at a somewhat accelerated rate, and in the third,

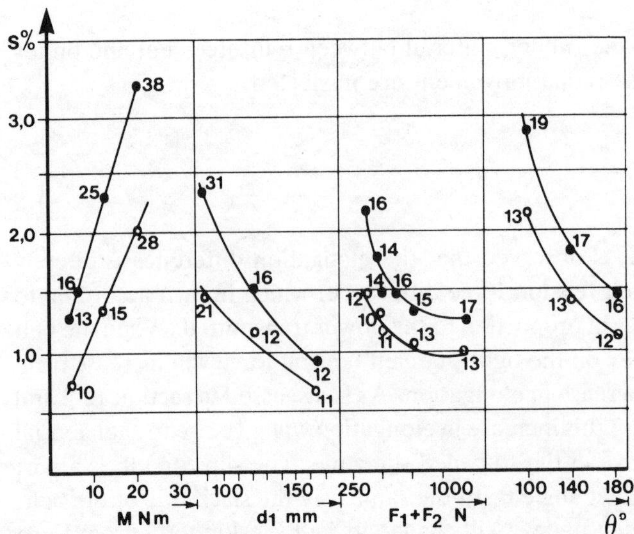

Figure 17-1. *Experimentally determined data for the dependence of slip s on torque transmitted M, small pulley diameter d_1, total belt tension F_1+F_2 and arc of contact angle θ. The experiments were carried out with classical V-belt of B profile with (●) and without (○) textile wrapping. The cut V-belts had cogged underside. Figures on the curves show the temperature rise (°C) in the belts when running (Breig and Oliver, 1980)*

at a fast increasing rate due to sliding slip. The transformation from region two to region three can be calculated and used as a measure of the power transmission capability of the belt.

Tension slip will lead to lower speeds in the driven pulley than would be expected from equation 17:1. If this lower angular velocity of the driven pulley is called ω_{s2}, efficiency in speed transmission η_ω and slip s we obtain:

$$\omega_{s2}=\omega_2 \cdot \eta_\omega=\omega_2 (1-s) \quad\quad\quad\quad (17:3)$$

Combining with equation 17:1 and using c_{ps} to denote speed ratio considering slippage we obtain:

$$c_{ps} = \frac{\omega_1}{\omega_{s2}} = \frac{\omega_1}{\omega_2 \cdot \eta_\omega} = \frac{\omega_1}{\omega_2 (1-s)} = \frac{c_p}{\eta_\omega} \quad\quad\quad\quad (17:4)$$

The equation shows that increased slip and elongation difference of the belt will increase speed ratio which sometimes must be considered when selecting standardized pulley diameters, cf section 18.6.

210

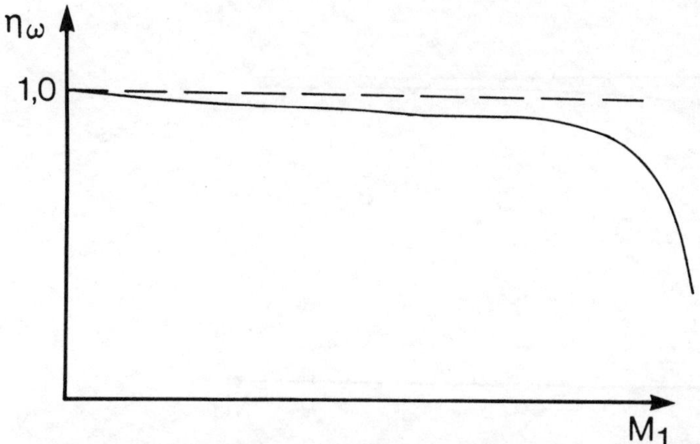

Figure 17-2. Efficiency of speed transmission η_ω as a function of torque transmitted M_1

Figure 17-2 shows how efficiency of speed transmission in principle decreases with increases in loading, here expressed as torque. The speed transmission efficiency is usually high, but will decrease rapidly with increases in slip due to overloads or large torques to be transmitted.

17.3 Efficiency in torque transmission

For the torque transmitted considering slip M_{s2} an efficiency of torque transmission η_M can, in the same way, be defined according to:

$$\eta_M = \frac{M_{s2}}{M_1} \cdot \frac{1}{c_p} \quad\quad\quad (17:5)$$

The torque transmission efficiency η_M equals zero, when no torque is transmitted, whereby the relation between η_M and torque M, in principle, can be graphed as in figure 17-3. The reason for η_M being smaller than 1 lies in friction and damping losses in the belt and its radial movement to be discussed in section 17.5. It should be noted that efficiency is a relative value given in relation to the torque transmitted. The decrease in torque transmission efficiency with reduced torque does not mean that losses increase – the case is rather the opposite one – but only that their relative share increases. A certain torque is necessary when the belt is running idle. In figure 17-3 the curve therefore starts at a certain value of M_1 denoting idle torque.

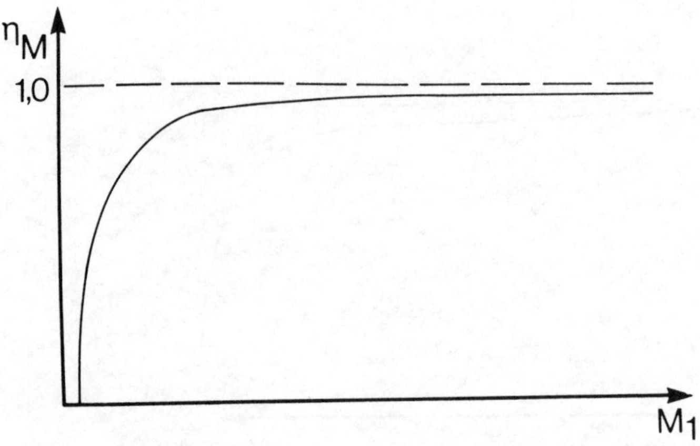

Figure 17-3. Efficiency of torque transmission η_M as a function of torque transmitted M_1

17.4 Efficiency in power transmission (total efficiency)

The total efficiency η_t expresses the efficiency in power transmission and is defined as the ratio between absorbed power P_2 on the driven shaft and power delivered P_1 from the driving shaft, i e:

$$\eta_t = \frac{P_2}{P_1} \quad\text{..}\quad (17:6)$$

The ratio can also be written:

$$\frac{P_2}{P_1} = \frac{M_{s2}\,\omega_{s2}}{M_1\,\omega_1} \quad\text{...}\quad (17:7)$$

Combining with equations 17:4 and 17:5, the following is obtained:

$$\eta_t = \eta_M\,\eta_\omega \quad\text{...}\quad (17:8)$$

The relation between total efficiency η_t and the torque transmitted is shown schematically in figure 17-4. When the belt drive is running idle, the total efficiency η_t by definition equals 0, but then increases rapidly with the load and levels out at a maximum level, which for most V-belt drives is found within

212

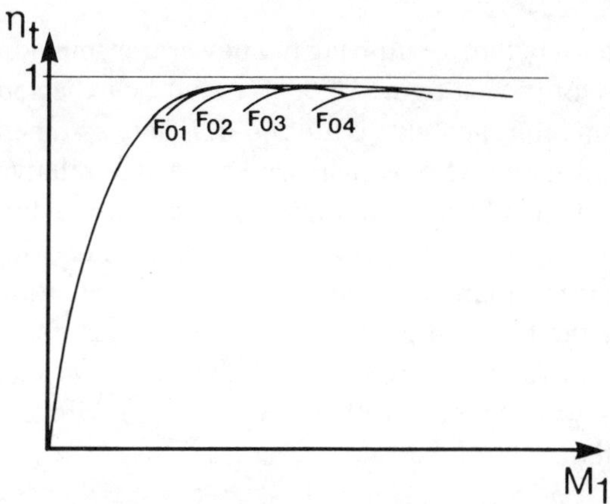

Figure 17-4. Total efficiency η_t as a function of torque transmitted M_1 at various levels of shaft load F_0 ($F_{01}<F_{02}<F_{03}<F_{04}$)

the range 92–97%, cf figure 4-7. Additional load increases will then only lead to excessive slip and a corresponding reduction in efficiency.

In figure 17-4 curves are given for various levels of shaft load F_0 obtained by various degrees of pre-tensioning. This shows that a maximum in the total efficiency is obtained at a given torque M and is related to the shaft loading F_0. The envelope curve gives the maximum efficiency that can be achieved by proper choice of pre-tension. Automatic regulation of pre-tension in proportion to the torque to be transmitted, cf section 14.7, will thus make optimum efficiency possible. Belts which transmit power independent of pre-tension, e g synchronous belts, show a more level curve when efficiency is plotted against torque.

One way of characterizing the efficiency is to describe losses as the sum of no-load losses and load-dependent losses. No-load losses are those typical of a given belt drive which are independent of load or power transmitted. If the no-load losses are denoted by $\triangle M_1$ and the load-dependent torque transmission efficiency by η_{Mb}, then equation 17:8 can be rewritten as:

$$\eta_t = \eta_\omega \, \eta_{Mb}\left(1 - \frac{\triangle M_1}{M_1}\right) \quad\quad\quad\quad\quad\quad (17:9)$$

The total efficiency is dominated by no-load losses, which normally make up 50–75% of the total losses. They usually comprise bending losses in the belt and bearing losses. Hereby the utilization of the power transmission capability will be the decisive factor. If thus a belt drive utilizing full power transmission capability has a total efficiency of 95% a figure of 90% would be expected when only half the power transmission capability is utilized. It is here assumed that the absolute value of no-load losses $\triangle M_1$ is unchanged but that its relative value is doubled. Experimental data published confirm this general trend, but usually total efficiency will be reduced at a somewhat slower rate with reductions in the power transmitted than could be predicted from equation 17:9. This can be explained by the fact that efficiency of speed transmission η_ω and efficiency of torque transmission η_{Mb} are somewhat dependent on load and that corresponding losses will decrease (efficiency increase) with decreases in the power transmitted.

17.5 Sources of energy losses

Energy losses can usually be described as friction losses, either caused by external, sliding friction or by internal friction. Some examples are:

- Bearing friction losses at the driving and driven shaft bearings as well as friction losses at shaft seals.
- Air friction losses during the motion of belts and the turbulence so created.
- Friction losses due to the slip of belts when leaving and entering the pulley grooves. These losses are characteristic for V-belts and are of less importance in flat belts.
- Frictional losses between belt and pulley groove caused by elongation slip and sliding slip. These losses will increase with transmitted power and mostly decrease with increases in arc of contact angle.
- Internal friction in belts during their bending over pulleys. These losses will increase with bending stress, i e with increases in belt height or decreases in pulley diameter. At a given belt speed they are independent of the power transmitted. Transmission belts with low belt height, e g flat belts or V-ribbed belts, give lower bending losses than V-belts. Cogged, cut V-belts give lower bending losses than uncogged, wrapped V-belts.
- Internal friction due to the increases and decreases in elongation between tight and slack side of the belt and pulsating side forces on the belt when it moves between the pulley grooves and the free span. The internal friction (damping) in the reinforcement material is the decisive factor.
- Additional friction losses caused by angular displacement or wear of pulleys.

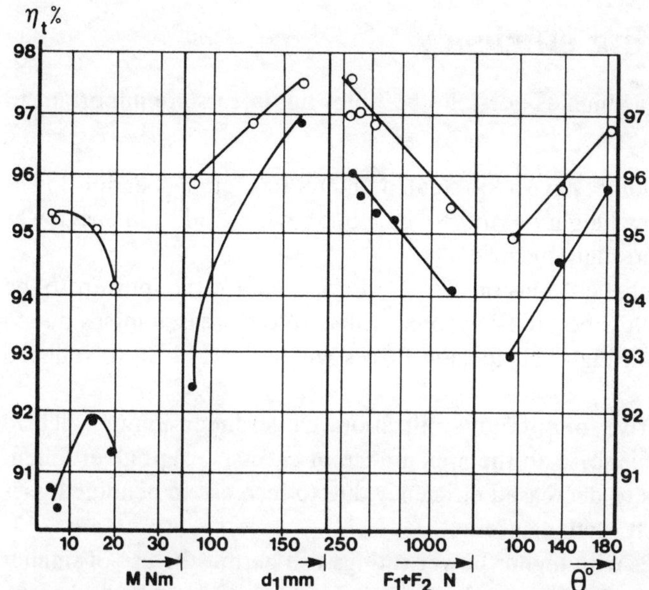

Figure 17-5. Examples of experimentally determined data for the dependence of total efficiency η_t on torque transmitted M, small pulley diameter d_1, total belt tension F_1+F_2 and arc of contact angle θ. The experiments were carried out with classical V-belts with B profile with (●) and without (○) textile wrapping. The cut V-belts had cogged underside, cf figure 17-1 (Breig and Oliver, 1980)

The combined effect of all these factors can be difficult to anticipate. Some published experimental data demonstrating the relation between total efficiency and various conditions in the belt drive are shown in figure 17-5. The dependence of efficiency on power transmitted mostly follows the trend illustrated in figure 17-4, but many deviations can be observed, which may be explained by experimental conditions used and the often rather limited interval over which measurements were carried out. Increase in small pulley diameter and arc of contact will generally influence efficiency in a positive direction. Increase in belt tension will have a positive effect if there is risk for excessive slip, but unnecessarily high belt tension will have a negative influence. The discussion above shows that efficiency is dependent on many factors and will vary with the conditions selected. Published data on the efficiency of various belt types often takes advantage of this fact by selecting conditions that will unduly favour a certain product.

17.6 Improving efficiency

To keep efficiency as high as possible the following factors are important to consider:

- Use pulleys that do not have a too small diameter to keep the bending losses low. Belts with low height or with cogging on the underside will give lower bending losses and higher efficiency.
- Select the proper belt tension force in relation to the power to be transmitted. Too low belt tension force will lead to increased losses due to excessive slip. Too high belt tension forces will similarly lead to reduced efficiency.
- Design the belt drive for optimum utilization. Underdimensioning will lead to decreased efficiency with the appearance of excessive slip. Overdimensioning also leads to decreased efficiency due to increase in bending losses in the heavier belt sections used.
- Modern belt types with higher power ratings will permit the use of smaller belt sections or of a smaller number of belts, cf figure 6-1, and will thereby give lower losses and higher efficiency.
- Defective, worn or improperly mounted bearings or pulleys can contribute to high energy losses.

18. Common design principles for V-belt drives with design example

18.1 Introduction

In this chapter the common procedure for belt drive design calculations for simple belt drives will be described and exemplified. A simple V-belt drive here means a belt drive with one driving and one driven pulley operating in the same plane. The procedure for design calculation is used and wellknown since long. It is recommended in many national standards, see for instance the references in Appendix C. Catalogues from V-belt manufacturers usually also contain details of this design calculation procedure, including the relevant graphs, charts, tables and coefficients necessary for the calculations. Data may differ somewhat from standards, mainly because catalogue data is compiled and fitted to a special product. Continued development leads to ever improved types of V-belts, e g with improved power rating or flexibility, to which catalogue data can be better matched than data in standards. Data in standards require a national or an international consensus which usually requires a long time before final acceptance is reached.

The design calculations are based on simplifying assumptions, but give acceptable accuracy for most practical applications. They have the further advantage of being simple to use. Many V-belt manufacturers have developed computer programs, also for use in minicomputers, which can additionally facilitate and speed up the calculations. The design principles discussed below are for V-belts, but may with some modifications also be used for flat belts, V-ribbed belts or synchronous belts. For more detailed information on design calculations for these belt types reference is made to relevant standards, cf the table in Appendix C, or to belt manufacturers' catalogues. For choice of belt pre-tension see chapter 14.

18.2 Starting data and aims for the design calculation

In design calculation of a V-belt drive the starting point is the following set of data, which must therefore be known:

- Type of driving machine and its power rating
- Type of driven machine
- Working hours of the drive
- Rotation speed of the fast-moving shaft
- Rotation speed of the slow-moving shaft
- Service conditions
- Approximate desired shaft centre distance

Starting from this data the design calculation is expected to give the following results:

- Belt type and belt section to be used
- Number of belts (multiple belt drive)
- Belt length, preferably a standard belt length
- Pulleys to be used, preferably standard pulleys (groove profile, number of grooves, diameter and shaft connection)
- Nominal shaft centre distance
- Installation and service allowances for the shaft centre distance

Design example:

- Driving machine: electric motor, 3-phase squirrel-cage motor, 4400 rpm (revolutions per minute), 24 kW
- Driven machine: centrifugal blower, approximately 2100 rpm, operating time 18 h daily
- Desired shaft centre distance: approximately 525 mm
- Service conditions: dirty environment (dust and sand)

Starting from this information, design data for the V-belt drive shall now be calculated.

18.3 Steps in the calculation procedure

The steps in the calculation procedure are given in figure 18-1.

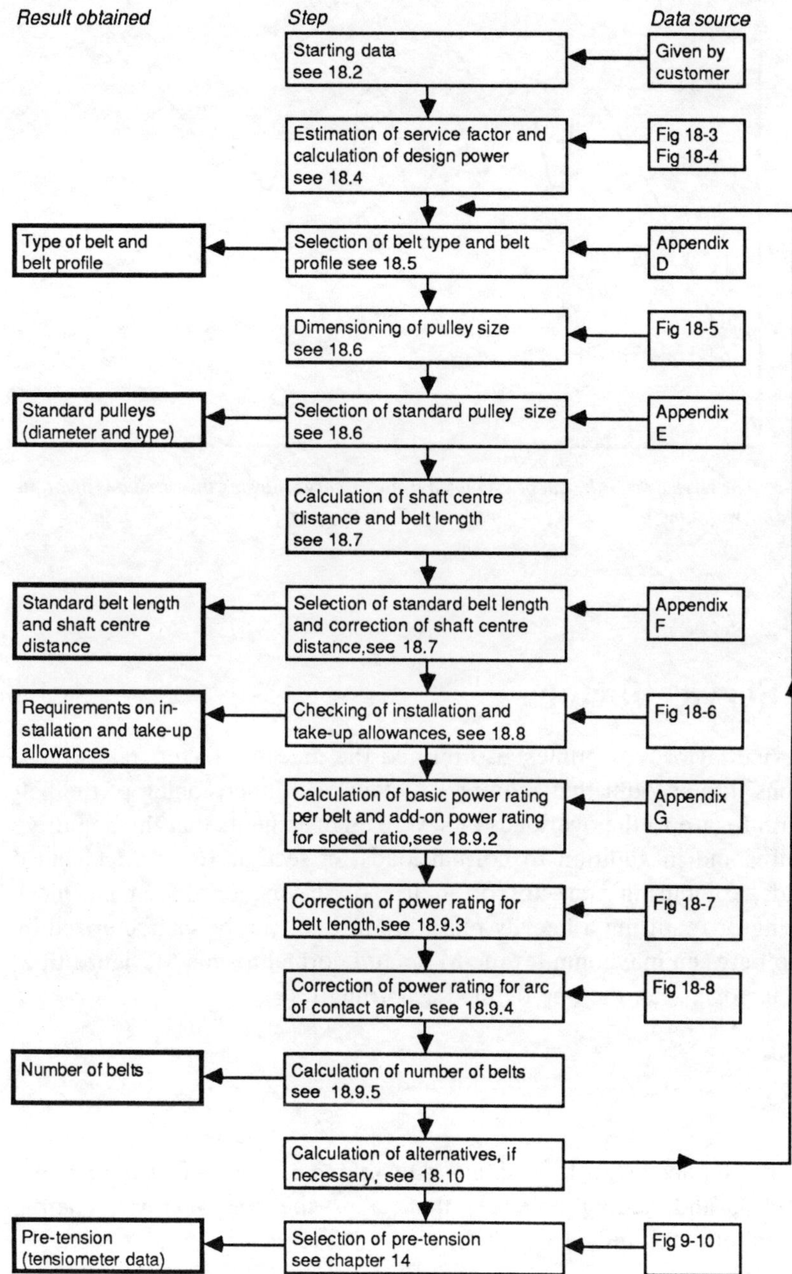

Figure 18-1. Scheme over the steps in design calculations of V-belt drives

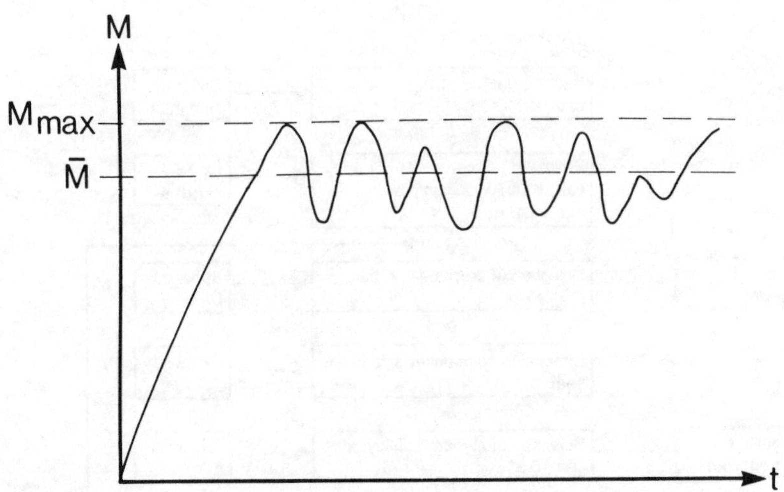

Figure 18-2. The service factor k_1 can be estimated from the curve showing the variations in torque transmitted M with time t

18.4 Service factor

The service factor, sometimes also termed the machine factor, takes those conditions into account that require a certain overdimensioning of the belt drive with regard to the overloads or severe environments that the belt drive must withstand in addition to normal loads, cf section 16.10. Mechanical overloads, e g during start-ups or in use of driving or driven machines generating or requiring a heavily pulsating torque, can be characterized by the ratio between maximum torque M_{max} and normal torque $\bar{M}$, figure 18-2. Service factor k_1 can then be defined according to:

$$k_1 = \frac{M_{max}}{\bar{M}} \quad \text{..} \quad (18:1)$$

Usually the service factor k_1 is obtained by taking a basic value from a table, figure 18-3, and adding a value, that takes the special environmental conditions into account, figure 18-4.

Design example:

Service factor $k_1 = 1,4 + 0,1 = 1,5$

220

Type of driven machine (type of drive	Type of driving machine					
	AC motors: Normal torque, one- or three-phase, starting torque approx. 150% of full load. DC motors: Shunt wound. Combustion engines: Multiple cylinder above 600 rpm.			AC motors: High torque, one- or three-phase, starting torque approx. 200–250% of full load. DC motors: Series wound, compound wound. Combustion engines: Single cylinder or multiple cylinder below 600 rpm.		
	Operating hours per day			Operating hours per day		
	<10	10–16	>16	<10	10–16	>16
Light duty	1,0	1,1	1,2	1,1	1,2	1,3
Medium duty	1,1	1,2	1,3	1,2	1,3	1,4
Heavy duty	1,2	1,3	1,4	1,4	1,5	1,6
Extra heavy duty	1,3	1,4	1,5	1,5	1,6	1,8

Light duty

Agitators for liquids. Blowers and exhausters. Centrifugal pumps and compressors. Fans up to 7,5 kW. Light duty conveyors.

Medium duty

Belt conveyors for sand, grain, etc. Dough mixers. Fans over 7,5 kW. Generators. Laundry machinery. Machine tools. Punches, pressés, shears. Printing machinery. Positive displacement rotary pumps. Revolving and vibrating screens.

Heavy duty

Brick machinery. Bucket elevators. Piston compressors. Drag, pan or screw conveyors. Hammer mills. Paper mill beaters. Piston pumps. Positive displacement blowers. Pulverizers. Saw mill and wood working machinery. Textile machinery.

Extra heavy duty

Chrushers. Ball or rod tube mills. Hoists. Rubber or plastic calenders, mills or extruders.

Figure 18-3. Examples of service factor in some practical applications (data from catalogue for TrellPower X V-belts from Trelleborg AB)

Internal idler pulley on slack side	0
External idler pulley on slack side	+0,1
Internal idler pulley on tight side	+0,1
External idler pulley on tight side	+0,2
Dirty environment (sand, dust, etc)	+0,1
Step-up speed ratio	+0,1
Temperatures up to 60°C	0
60–80°C	+0,1
80–100°C	+0,2
Wet conditions involving fluids that do not affect the belt chemically	+0,1
Reversing or shock loadings – please contact the supplier	

Figure 18-4. Examples of recommended add-on figures to the service factor to take care of adverse conditions (from catalogue for TrellPower X V-belts from Trelleborg AB)

The value 1,4 is valid for a 3-phase, squirrel-cage electric motor operating more than 18 h daily according to figure 18-3. The added value 0,1 holds for dirty conditions (dust and sand) according to figure 18-4.

The design power P_d of the belt drive is now obtained by multiplying the motor power P_m with the service factor k_1 according to:

$$P_d = k_1\, P_m \quad\quad\quad\quad (18{:}2)$$

Design example:

$$P_d = 1{,}5 \cdot 24 = 36 \text{ kW}$$

18.5 Selection of belt type and belt profile

The choice of belt type is important but in most cases self-evident. Narrow V-belts are a modern development of classical V-belts and cut V-belts can in most cases be looked upon as improvements over wrapped V-belts. In our design example it is therefore natural to choose a cut V-belt with a narrow belt profile. The choice of suitable belt profile is usually made with the aid of graphs, relating design power P_d to rpm of the smaller pulley using double-logarithmic plotting. These diagrams show the useful areas for various belt sections, cf example in Appendix D.

Design example:

For a design power of 36 kW and a rotation speed n_1 of the small pulley 4400 rpm the chart in Appendix D shows that either a narrow V-belt section SPZX or a classical V-belt section AX should be used. The letter "X" is used here, as by many V-belt manufacturers, to denote cut belts. The modern type SPZX is selected.

If drive conditions fall on the borderline between two belt profiles in the chart, the smaller profile is usually selected, which will decrease the cost and space requirements of the drive.

18.6 Choice of pulleys

The type of pulley to be used is usually determined from experience and by the relevant service conditions. High power ratings, long daily operating hours and severe operating conditions usually require pulleys of the best possible manufacture and materials, usually made from cast iron or steel. The choice between pulleys with integral hubs or pulleys with dismountable bushings is according to section 5.2 usually made with regard to demand for easy installation. At low power ratings, intermittent service and not too severe operating conditions cheaper and lighter pulleys may be used, for instance made from aluminium or plastics. It should be noted, however, that the increased power ratings obtained with modern V-belts increase the demands on the quality of the pulleys.

The diameter of the small pulley is calculated from the belt speed that can be permitted. According to section 16.6 power rating will increase with belt speed up to the point where the effect of centrifugal forces will become predominant. Starting from a given belt speed, a small pulley diameter is read off from the chart in figure 18-5. When this has been done the closest standard pulley diameter is selected, for instance from the table in Appendix E. With the aid of the speed ratio, equation 17:1, the large pulley diameter d_2 can now be calculated. The closest standard pulley diameter is then selected, for instance by using the table in Appendix E. Certain deviations from the desired speed ratio must then be tolerated. If a more exact speed ratio is required the most economical solution will usually be to select a large pulley of standard dimensions while the small one is specially manufactured to order. If the design calculations for the large pulley will give a diameter between two standard diameters the smaller diameter should be selected to compensate for slippage, cf section 17.2.

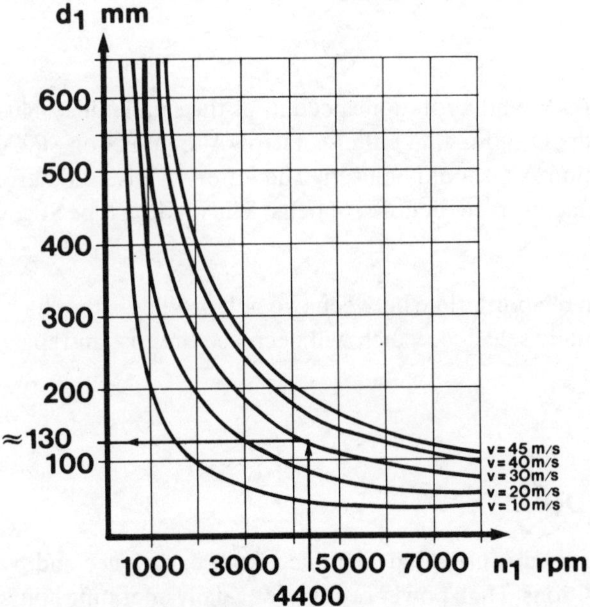

Figure 18-5. Chart for selection of small pulley diameter d_1 starting from desired rotation speed n_1 and permissible belt speed v (data for the designed example is indicated)

Design example:

Starting from a belt speed of 30 m/s and a rotation speed of 4400 rpm figure 18-5 shows that a small pulley diameter of 130 mm should be used. The table in Appendix E gives the closest standard diameter as 132 mm. By insertion in equation 17:1:

$$c_p = \frac{d_2}{d_1} = \frac{n_1}{n_2} = \frac{d_2}{132} = \frac{4400}{2100} = 2,10$$

This will give $d_2 = 277$ mm. From the table in Appendix E the closest standard pulley diameter is found to be 280 mm, which will give the desired speed ratio with an acceptable accuracy.

18.7 Calculation of shaft centre distance and belt length

Two cases may be at hand: either the choice of shaft centre distance is free or a given desired shaft centre distance is prescribed. In the first case shaft

224

centre distance c is taken as the largest value of either c_1 or c_2, defined according to:

$$c_1 = \frac{d_2 + 3\,d_1}{2} \quad\text{...}\quad (18{:}3)$$

$$c_2 = d_2 \quad\text{...}\quad (18{:}4)$$

From the shaft centre distance belt length can then be calculated with the aid of equation 11:8. In the second case, i e when a given shaft centre distance is desired, belt length is calculated according to equation 11:8 by inserting this value. The closest standard belt length is then chosen from tables, cf example in Appendix F. To fit a standard belt length, the shaft centre distance must then be corrected somewhat. This is usually done by increasing or decreasing the shaft centre distance with half the difference between the standard belt length and the belt length originally calculated, i e by application of equation 11:8 in a simplified form.

Design example:

A shaft centre distance of 525 mm is desired. Pulley diameters were selected to 132 and 280 mm respectively. Belt length L is now calculated with the aid of equation 11:8:

$$L = 2c + \frac{\pi}{2}(d_2 + d_1) + \frac{(d_2 - d_1)^2}{4c} = 2 \cdot 525 + \frac{3{,}14}{2}(280 + 132) +$$

$$+ \frac{(280 - 132)^2}{4 \cdot 525} = 1707 \text{ mm}$$

The table in Appendix F shows that the closest standard belt length for belt section SPZX is 1700 mm. The difference between the standard length and the originally calculated length is 7 mm, why the originally calculated shaft centre distance 525 mm is adjusted by subtracting half this difference, i e 3 mm, giving a final shaft centre distance of 522 mm.

If no requirement on shaft centre distance were given, c_1 and c_2 can be calculated according to equations 18:3 and 18:4 giving:

$$c_1 = \frac{d_2 + 3\,d_1}{2} = \frac{280 + 3 \cdot 132}{2} = 338 \text{ mm}$$

$$c_2 = d_2 = 280 \text{ mm}$$

The largest of these two numbers is 338 mm, which is then selected as the shaft centre distance, after which belt length is calculated as above and the shaft centre distance adjusted according to the standard belt length chosen.

18.8 Installation and take-up allowances

As V-belts have certain belt length tolerances and will furthermore be elongated when installed and creep and relax during service, shaft centre distances must be adjustable to a certain degree to allow for tensioning of the belts during installation and readjustment of tension in service. The adjustment possibilities will also permit easy installation of the belts without prying them into the grooves. The allowances needed will increase with the length and section size of the belt, figure 18-6.

Pitch length, mm	Installation and take-up allowances, mm				
	−				+
	SPZX/ 3VX	SPAX	SPBX/ 5VX		all profiles
612–1180	15	20	—		35
1181–1800	20	25	25		35
1801–2650	20	25	25		40
2651–3150	20	25	25		45
3151–3350	20	25	25		60

Pitch length, mm	Installation and take-up allowances, mm				
	−				+
	ZX	AX	BX	CX	all profiles
604–700	15	20	25	—	25
701–1000	15	20	25	—	25
1001–1580	20	20	25	40	30
1581–2350	20	20	30	40	40
2351–3100	—	20	30	40	50
3101–3379	—	25	30	40	65

Figure 18-6. Examples of recommended installation and take-up allowances for V-belt drives. SPZX/3 VX, SPAX and SPBX/5 VX are narrow, cut V-belt profiles and ZX, AX, BX and CX classical, cut V-belt profiles (data for TrellPower X V-belts in catalogue from Trelleborg AB)

Design example:

With belt section SPZX and belt length 1700 mm the installation and take-up allowance should be $+35/-20$ mm according to figure 18-6. Shaft centre distance should consequently be adjustable between $522-20=502$ mm and $522+35=557$ mm.

18.9 Power rating per belt and number of belts

18.9.1 Principle of calculation

The number of V-belts Z in the drive is obtained by dividing design power P_d of the drive, calculated according to equation 18:2, by the power that can be transmitted per belt according to:

$$Z = \frac{P_d}{(P_b + \triangle P_{b1})\, k_2\, k_3} \quad\dots\dots\dots\dots\dots\dots\dots\dots\dots\dots\dots\dots\dots\dots (18:5)$$

To the basic power rating P_b per belt is thus first added an add-on power rating $\triangle P_{b1}$ for speed ratio, after which the sum is multiplied by a belt length correction factor k_2 and an arc of contact correction factor k_3.

18.9.2 Basic power rating and add-on power rating for speed ratio

Power rating of a V-belt with given dimensions and construction is mainly dependent on belt speed and small pulley diameter. Equations can be set up for the calculation, cf section 16.11, but values are usually obtained from tables. Tables that are valid for a given type and profile of belt, give the basic power rating per belt as a function of the small pulley diameter, which is a measure of flexural stresses, and the rpm of the small pulley, which is a measure of belt speed. The same table usually gives figures for both basic power rating P_b and add-on power rating for speed ratio $\triangle P_{b1}$. Figures for the add-on power rating for speed ratio are also given for various rpm's of the small pulley, cf table in Appendix G.

Design example:

Starting from a small pulley with diameter 132 mm and running at 4400 rpm the table in Appendix G gives for belt section SPZX a basic power rating per belt $P_b = 9,86$ kW and an add-on power rating for speed ratio $\triangle P_{b1} = 0,47$ kW.

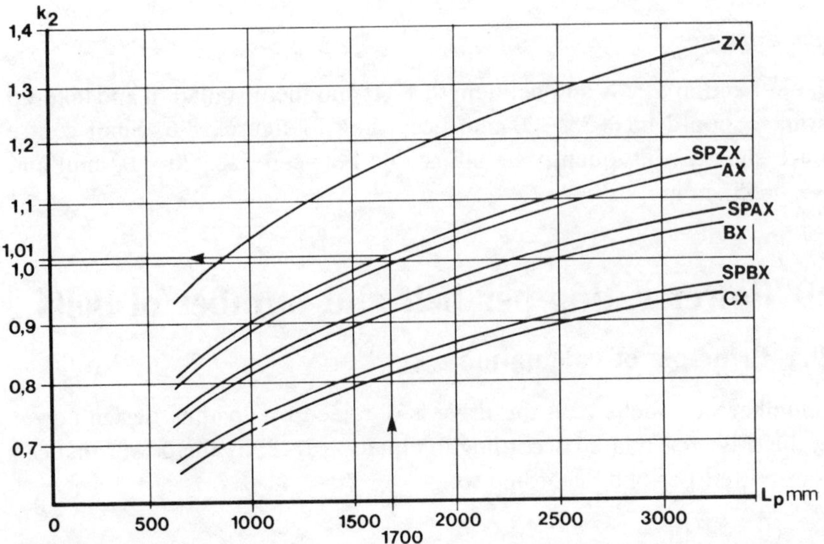

Figure 18-7. Correction factor k_2 for belt length as a function of pitch length L_p for different profiles of cut V-belts. Data for the design example is indicated (from catalogue for TrellPower X V-belts from Trelleborg AB)

18.9.3 Belt length correction factor

With increased belt length the number of flexures and number of tension stress changes between slack and tight side of the belt per unit of power transmitted will be reduced, meaning lowered belt operating temperature and prolonged belt life, cf section 16.9. While retaining belt life a higher power rating can thus be permitted, which is taken into account by multiplication with a belt length correction factor k_2 according to equation 18:5. The basic power ratings and add-on power ratings given in tables, cf example in Appendix G, are referred to a base length, which is different for each belt section and increases with its size. At the base length k_2 will thus be 1, at lengths shorter than the base length <1 and at belt lengths larger than the base length >1. This dependence can be tabulated or given in graphs, figure 18-7. It shows that k_2 normally varies within the interval $0,65-1,3$. Belt length can also be considered by introducing an add-on power $\triangle P_{b2}$, cf equations 16:16 and 16:20.

Design example:

For belt section SPZX with belt length 1700 mm a value of $k_2=1,01$ can be read off from figure 18-7.

228

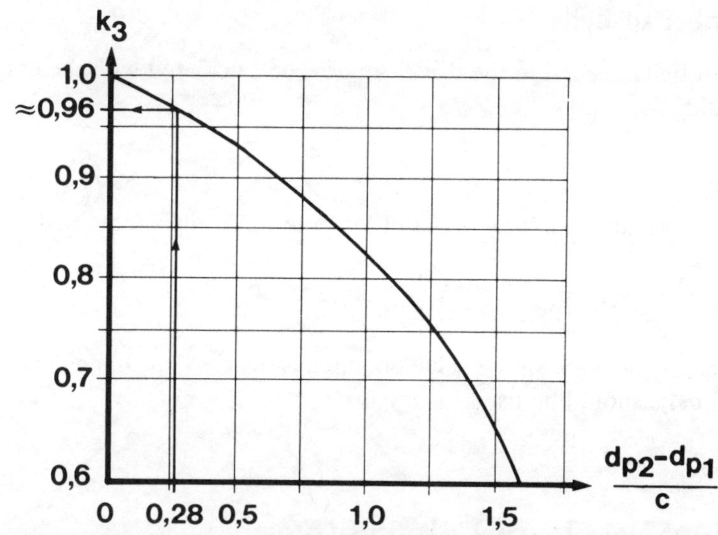

Figure 18-8. Correction factor k_3 for arc of contact angle. As a measure of arc of contact angle is taken the ratio between the difference in pulley diameters $d_{p2}-d_{p1}$ and shaft centre distance c. Data for the design example is indicated

18.9.4 Arc of contact angle correction factor

Power rating capability will decrease when arc of contact angle θ is decreased. Tabulated values of basic power rating and add-on power rating for speed ratio, e g according to Appendix G, are given for an arc of contact angle $\theta=180°$. For lower values of θ on the small pulley the sum of base power rating and add-on power rating must be multiplied with an arc of contact angle correction factor k_3. This factor is 1 for $\theta=180°$ and will decrease down to 0,65 for $\theta=83°$. A mathematical expression for the relation between power rating and arc of contact angle is given by Eytelwein's equation 12:16, cf also equation 16:18. To simplify calculations tables or charts are usually used, cf figure 18-8. The arc of contact angle need not be calculated if the correction factor k_3 is given as a function of the ratio between pulley diameter difference and the shaft centre distance utilizing equation 11:5 or 11:6.

Design example:

Pulley diameter $d_1=132$ mm, $d_2=280$ mm and shaft centre distance c=522 mm. From this is calculated $(d_2-d_1)/c=0,28$, which using the graph in figure 18-8 gives $k_3=0,96$.

18.9.5 Number of belts

The number of belts needed in the drive can now be calculated with the aid of equation 18:5.

Design example:

By inserting actual numbers into equation 18:5 the following is obtained:

$$Z=\frac{P_d}{(P_b+\triangle P_{b1})\ k_2\ k_3}=\frac{36}{(9{,}86+0{,}47)\ 1{,}01\ \cdot\ 0{,}96}=3{,}6$$

Usually the integer next above the value obtained is selected, meaning in this case that 4 V-belts should be used for the drive.

18.10 Final result and alternatives

The final result of the design calculation is that a small pulley with a standard diameter of 132 mm and a large pulley with a standard diameter of 280 mm should be selected, which gives a speed ratio $c_p=2{,}12$. The rpm of the driven shaft will then be 2074 rpm against desired 2100 rpm. In service an additionally reduced rpm is probably to be expected due to elongation slip in the drive. The design calculation also shows that the required power can be transmitted by use of 4 cut narrow V-belts of profile SPZX. The standard belt length of 1700 mm selected corresponds to a shaft centre distance of 522 mm, very close to the originally desired value of 525 mm. Installation and service allowances on shaft centre distance should be at least +35/−20 mm. Pulley bushings, locking devices and centre bores should be selected to fit on the shafts.

The design calculation, as carried out above, usually leads to the best technical solution, unless special conditions are to be considered. As the calculations are so easy and rapid to perform – especially if they are made with the aid of computers or minicomputers – alternative calculations, based on various assumptions, can be valuable and interesting, and can give economically attractive alternatives. Below a few examples will be presented on alternative calculations, starting with the data given:

Alternative 1:

In choice of pulley diameters above a belt speed of 30 m/s was used. Cut, narrow V-belts may, however, be used up to belt speeds of 45 m/s. With this belt speed small pulley standard diameter $d_1=200$ mm and large pulley standard diameter $d_2=400$ mm can be chosen, giving a speed ratio $c_p=2{,}00$. The necessary power can be transmitted with 3 (actually 2,4) cut, narrow V-belts of profile section SPZX. With a standard belt length of 2025 mm shaft centre distance will be $c=532$ mm. The calculation thus shows that by utilizing the maximum allowable belt speed the number of belts in the drive can be reduced from 4 to 3. The benefit gained is doubtful, however, as larger and heavier pulleys have to be used, which require more space. The high speed allowed with modern V-belt types is usually taken to advantage as high motor rpm's are increasingly used. Lighter and cheaper electric motors may then be used for a desired power to be transmitted.

Alternative 2:

Another way of reducing the number of belts in the drive is by use of a heavier belt section, e g SPAX. With belt speed 30 m/s standard pulley diameters will be the same, i e 132 mm for the small pulley and 280 mm for the large pulley, giving a speed ratio $c_p=2{,}12$. The power required can then be transmitted with the aid of 3 (actually 2,8) cut narrow V-belts of belt profile SPAX. With a standard belt length of 1700 mm shaft centre distance will be $c=521$ mm. This may be a good alternative to the original design calculation.

Alternative 3:

If we want to use cut, classical V-belts for the drive, profile AX is the most suitable one. Also in this case, the standard pulley diameters will be the same, i e $d_1=132$ mm and $d_2=280$ mm, giving a speed ratio $c_p=2{,}12$. To transmit the desired power 5 (actually 4,9) cut, classical V-belts of profile AX will be required. With a standard belt length of 1697 mm, the shaft centre distance will be $c=520$ mm. The larger number of belts required will probably make this drive less economical compared to the cut, narrow V-belts. The selected belt speed of 30 m/s is also somewhat too high for classical V-belts. If a belt speed of 20 m/s is chosen as more reasonable, the standard diameter of the small pulley will become $d_1=85$ mm and of the large pulley $d_2=180$ mm, giving a speed ratio $c_p=2{,}12$. To transmit the desired power 9 (actually 8,2) cut, classical V-belts of profile AX are required. With a standard belt length of 1454 mm a shaft centre distance of 517 mm will be needed. The large number of belts required will make this design unwieldy and is mainly a result of the small pulleys used that give increased bending stresses leading to a reduced life or alternatively, lower power ratings.

Alternative 4:

In order to reduce the number of classical V-belts the larger profile BX of cut, classical V-belts is chosen. Starting from a belt speed of 30 m/s pulley diameters will be the same, i e $d_1=132$ mm and $d_2=280$ mm, giving a speed ratio $c_p=2,12$. To transmit the required power 4 (actually 4,0) cut, classical V-belts of profile BX will be required. With a standard belt length of 1692 mm shaft centre distance will be $c=517$ mm. Also in this case it may be remarked that the belt speed is too high. Starting from the more reasonable belt speed of 20 m/s, the small pulley diameter d_1 will become 85 mm. This figure is below the minimum pulley diameter recommended for profile BX, which is 118 mm. The lower speed can thus not be recommended as an alternative.

The examples given above show how design calculations can be carried out starting from a given set of data. Often conditions can be very restricted, for instance when pulleys of given diameters, groove profiles and number of grooves have to be used. If there is any doubt about the best alternative to use, V-belt manufacturers will always offer service and advice.

18.11 V-belt profiles for which design data are missing

From chapter 3 it can be seen that besides the normal, often well standardized V-belt types, there exists a great number of varieties. As the preparation of power rating tables requires much calculation work, the results of which have to be confirmed by experimental investigations, it may sometimes be difficult to find suitable design data for odd V-belt types or special varieties. Power ratings can sometimes be calculated with the use of specially developed equations, cf section 16.11. These equations usually require, however, that values of the various coefficients are known from experimental investigations. Another method is to compare with a standardized belt profile of approximately the same dimensions and with equivalent type and amount of reinforcement, and apply corresponding power ratings. Safety is achieved by overdimensioning, e g avoiding very small pulley diameters, very high belt speeds and high power ratings.

Hexagonal belts can be looked upon as two classical V-belts placed with their top sides against each other and with the same reinforcement as corresponding classical V-belt. Hexagonal belt profile HAA can thus be compared to classical V-belt profile A, HBB to B, etc. As hexagonal belts show no advantage in simple belt drives, the normal design calculation procedure is mostly without interest. If an estimation is desired the same power ratings as for corresponding classical V-belts can be used. A certain safety factor should be applied, however, as hexagonal belts have larger height than classical V-belts, leading to larger bending stresses, increased operating temperatures and a reduction in belt life. For belt drives with more than two pulleys, cf section 19.3, the same principles for design calculations are used as with narrow or classical V-belts. In calculations of fatigue rates and belt life, the actual value of the small pulley diameter is not used but instead 75% of it to compensate for increased bending stresses and operating temperature.

Joined V-belts may be looked upon as the same number of corresponding individual narrow or classical V-belt sections. The profile height in joined V-belts is usually somewhat larger than for corresponding narrow or classical V-belt sections to avoid contact between pulleys and the adjoining top layer. The adjoining layer will further increase bending stresses. This negative effect is compensated for, however, by the ability of the joined V-belt to give a more even load distribution between the individual V-belt sections. The end result will be that the same power rating and belt life can be expected as with the corresponding number of individual V-belt profiles. To this should be added the advantages mentioned in section 3.5 of joined V-belts to dampen vibrations and take up shock loads without risk for tilting.

18.12 Drive design with flat and V-ribbed belts in brief

The principles for design calculations of belt drives using flat belts or V-ribbed belts can be found in standards, e g the British standard BS 351 for flat belts or the American standard RMA IP–26 for V-ribbed belts, and in belt manufacturers' catalogues. Largely the same procedure is followed as given above for V-belts. The most important difference is that multiple drives are not used. Rather the width of the belt is increased to take high power ratings. Alternatively at a given width the thickness of the belt section can be varied. Power rating is therefore usually given per unit of width of the belt, e g per 25 mm of width of flat transmission belts and per rib for V-ribbed belts.

Sometimes design calculations start from a permitted force per unit width of the belt from which the power rating can be calculated with the aid of the actual width of the belt, its speed and the effective belt tension ratio. When using open-ended flat transmission belts that are joined by mechanical fasteners on site, no regard has to be taken to any standard belt lengths when determining the shaft centre distance.

18.13 Drive design with synchronous belts in brief

Principles used in design calculations of synchronous belt drives are given in standards, e g the American standard RMA IP–24 or the British standard BS 4548, and in belt manufacturers' catalogues. Largely the same procedure is used as given above for V-belts. Among the divergences the following can especially be noted:

1. Belt length and circumferences of pulleys must be integers of the pitch length p_b according to equations 11:13, 11:14 and 11:15.
2. Multiple belt drives are normally not used and the desired power rating is obtained by the choice of belt section and belt width. The width of the belt should not exceed the small pulley diameter to guarantee that it will keep track properly. Power rating is usually given per unit width of belt, for instance per 25 mm. It is assumed to increase faster than in direct proportion to belt width, which is usually expressed with a tabulated correction factor. As an example the correction factor may be 0,42 for belt width 12,5 mm, 1,00 for belt width 25 mm and 2,14 for belt width 50 mm. The reasons for the correction factor is to be found in the method of manufacture used for the belts. They are usually manufactured in large widths and the cord reinforcement applied by spiral-winding. In cutting up this belt into smaller widths the cord will be cut through at each cut. The corresponding reduction in force and power uptake will be relatively larger the smaller the width of the belt.
3. Power is transmitted by mechanical interlocking and not by friction as with V-belts. The decisive factor for the power rating will thus be the forces that the cogs in the belt can take up. The number of cogs (teeth) in mesh will have the same importance as the arc of contact for V-belts. If the number of cogs in mesh is less than 6 a correction factor must be applied, the value of which is tabulated.

4. Pulleys may need to be flanged on one or both sides to keep belts in place when running. The minimum requirement is one pulley flanged on both sides or both pulleys having one flange on opposite sides. If shaft centre distance is larger than 8 times the small pulley diameter both pulleys should be flanged on both sides. If the shafts are vertical, at least one pulley should be flanged on both sides and all the others at least on the underside.
5. Synchronous belts cannot, unlike V-belts or flat belts, be disengaged by slippage when overloaded. This is usually considered during design calculations by applying a higher service factor, usually within the range 1,5–2,5.

19. Design calculations for special drives

19.1 Introduction

It has been said that the general design calculation of V-belt drives, as given in chapter 18, covers more than 90% of all applications. As it only applies to belt drives with two pulleys operating in the same plane and is based on a working life of the drive of 25 000 h, other applications can be found, however, which require other methods of calculation. In this chapter some examples will first be given for cases involving slight modifications of the normal design procedure. Finally, more complicated cases will be presented, which require more advanced calculation methods, for which the computer is often an indispensable tool. At the end of the chapter calculation methods for vibrations in V-belts are briefly discussed.

19.2 V-belt drives with step-up speed ratio

The overwhelming number of V-belt drives are designed for step-down speed ratios, i e the rpm of the driving shaft is higher than that for the driven shaft. This means that the small diameter pulley is placed on the driving shaft and the large diameter pulley on the driven shaft. Exceptionally step-up speed ratios are used, meaning that the large pulley is placed on the driving shaft, which has lower rpm than the driven shaft, on which the small pulley is placed. Design calculations can then be performed with the standard procedure as long as it is observed that the small pulley is placed on the driven shaft. The belt speed should be checked so that it doesn't become excessively large.

19.3 Belt drives with idler pulleys

The use of idler pulleys to control belt tension is discussed in section 14.5. Idler pulleys do not transmit power, meaning that belt tension forces on the belt strands entering and leaving the idler pulley are equal. Idler pulleys will, however, cause extra flexure of the belt, meaning either reduced belt life or reduced power rating. Especially unfavourable are idler pulleys placed on the outside of the belt, as this leads to double-bending. This placement will on the other side have the advantage that the arc of contact on the small diameter pulley will be increased, leading to an increased value of the correction factor k_3 for arc of contact angle and a higher power rating, according to equation 18:5. The standard design calculation procedure can be used, with the following modifications:

1. Pulley diameters are calculated in the usual way.
2. Belt length is calculated with equation 11:8 inserting figures for pulley diameters and shaft centre distance, which is usually not adjustable when idler pulleys are used. A corrected belt length is obtained by adding 2 times the plus allowance for shaft centre distance according to figure 18-6 to the calculated belt length according to equation 11:8. This will allow easy installation of the belt on the pulleys. The closest standard belt length is then chosen. Combination of large minus-tolerances on shaft centre distance and large belt lengths can mean a risk that belt strands come into contact with each other with external placement of the idler pulleys.
3. The placement and diameter of the idler pulley are chosen according to section 14.5.
4. Calculate the position of the idler pulley with the aid of the pulley and idler pulley diameters, placement of pulley and shaft centre distance. This calculation will be relatively complicated, if accuracy is desired, and is preferably carried out with the aid of a computer. A drawing made to scale can also be used for estimating belt span length and belt contact length on pulleys and then by repeated fitting at least an approximately correct value is obtained. Belt span length and belt contact length on pulleys can also be calculated with the aid of equations 11:11 and 11:12. Creep and wear of the belt can influence the positioning of the idler pulley.
5. With a given position of the idler pulley, the arc of contact angle for the small pulley can be calculated, either manually and approximately with the aid of a scale model of the drive or more exactly with the aid of a computer program.
6. For calculating power rating and number of belts required the standard design calculation procedure is used. The extra bending stresses, caused by the idler pulley, are taken into account by using a somewhat larger service factor according to figure 18-4. Alternatively a correction factor k_4 for idler pulley can be introduced into equation 18:5, which may then be written:

$$Z=\frac{P_d}{(P_b+\triangle P_{b1})\ k_2\ k_3\ k_4} \quad\dotfill (19{:}1)$$

Typical values for k_4 are 0,91 for one idler pulley, 0,86 for the more unusual case with two idler pulleys and 0,81 for three idler pulleys. Both calculation methods will give a decrease of about 10% in power rating when using one idler pulley.

7. In selecting belt tension for the idler pulley, the great difference should be observed between the two cases when the idler pulley is mounted in a fixed position and where it is freely movable under the influence of weights or spring forces. Mounting in a fixed position is carried out according to the principles applied for belt drives without idler pulleys. If the idler pulley is freely movable and weight- or spring-loaded, the slack side tension can be kept at a fixed and low value independent of the power transmitted. Tight side tension will then be automatically regulated in proportion to the power to be transmitted. The end result will be that a weight- or spring-loaded idler pulley on the slack side will lead to a low average belt tension and thereby an increased life of belts and bearings. Take as an example a belt drive which requires a belt tension difference $F_{e1}-F_{e2}=100$ N to transmit the desired power. If $F_{e1}{:}F_{e2}$ is assumed to be 5:1 at full load, cf section 16.4, a belt pre-tension of 75 N corresponding to a bearing load of approximately 150 N is necessary when placing the idler pulley in a fixed position. At full load $F_{e1}=125$ N and $F_{e2}=25$ N is obtained and at half load $F_{e1}=100$ N and $F_{e2}=50$ N. If on the other hand the idler pulley is weight- or spring-loaded to give a slack side tension of $F_{e2}=25$ N, at full load F_{e1} will be $=125$ N and bearing load 150 N, while at half load $F_{e1}=75$ N and bearing load approximately 100 N, i e considerably lower average belt tension and lower bearing loads than when placing the idler pulley in a fixed position.

19.4 Quarter-turn and crossed belt drives

Figure 2-5 shows the drive arrangement for quarter-turn and crossed belt drives. Quarter-turn drives are used for transmitting power between pulleys and shafts in different planes. Crossed belt drives cause the reversing of rotation between driving and driven shaft. Simple mathematical expressions for calculation of belt length in quarter-turn drives are not available. Estimations or model trials may be used, alternatively calculations with the aid of a computer. Belt length for crossed belt drives can be calculated using equation 11:9.

In belt drives of this type great care has to be taken in the design to protect the belts from the effects of the torsional strain imposed. For crossed belt drives shaft centre distance should not fall below certain values, which are tabulated in some design manuals. Furthermore, it must be avoided that belt strands in the middle of the span rub against each other, which may cause rapid wear. Guiding rollers can be used to keep the strands apart. Crossed belt drives are not suitable for use as multiple belt drives. Quarter-turn drives should preferably be designed so that the belt enters the pulley plane at an angle not larger than 5–6°. This is reached by limiting speed ratio c_p to a maximum of 2,5:1 and to have a minimum shaft centre distance c, calculated from the large pulley diameter d_2 and width of the band of belts w_w according to:

$$c \geq k_5 \ (d_2 + w_w) \qquad\qquad\qquad\qquad\qquad\qquad\qquad (19:2)$$

The coefficient k_5 can for instance have the value 5,5 for shafts at 90° angle, 4,0 for 45° angle and 3,0 for 30° angle. Cogged belts and joined V-belts are usually not suited for crossed or quarter-turn belt drives. If larger speed ratio than 2,5:1 is desired in a quarter-turn belt drive, this can be obtained by using a drive with a jackshaft, from which the desired higher speed ratio is obtained in a drive with aligned pulleys. To avoid that belts run off the pulleys deep groove pulleys are preferred. In crossed belt drives pulleys are usually placed in the same plane. In quarter-turn drives the pulley on the vertical shaft should be displaced somewhat downwards towards the tight side of the belt, which should always be on the underside.

For selection of belt profile the same drive design calculation procedure and power rating tables are used as described in chapter 18. A somewhat shorter belt life or alternatively lower power rating is unavoidable, why it is recommended to apply a certain safety factor in the design calculations.

19.5 V-belt drives with one flat pulley

In some belt drives only the driving (small) pulley is grooved, while the driven (large) pulley has a flat, cylindrical surface. The arrangement is sometimes termed V-plane or V-flat drive as opposed to V-V-drive which indicates that both pulleys are grooved. These drives were earlier often used when changing from flat belts to V-belts. The large, flat pulley was then retained and the small, flat pulley was changed to a grooved pulley. A reason to still use this type of drive may be the desire to have a heavy pulley with flywheel effect. The drive will usually function well with classical V-belts, but narrow belts tend to tilt on the flat pulley and should be avoided. Only drives with horizontal shafts should be used. Shaft centre distance should not exceed the large pulley diameter if the pulley is not flanged. Acceptable power ratings require that $d_2 \geq 3\ d_1$. In V-belt drives the ratio between belt tension forces in tight and slack sides respectively, is chosen at $F_{e1}:F_{e2}=5:1$. In drives with one flat pulley a higher ratio than 2,5:1 cannot be recommended. This means that the correction factor k_3 for the arc of contact angle in equation 18:5, becomes somewhat smaller than the values suggested in the chart in figure 18-8. Tabulated values can be found in design manuals.

19.6 Design calculations with belt life as a variable

Design calculations and power ratings are usually based on a belt life of 25 000 h. This is the normal requirement in industrial belt drives and corresponds to a useful life of 3–5 years. As is discussed in section 15.8 shorter belt lives may be acceptable in many applications, especially in intermittent or seasonal service. More accurate calculation methods with belt life as a variable are discussed in section 19.7. In many cases simpler calculation methods can be used, e g:

1. Some V-belt manufacturers have tabulated power ratings for different values of belt life, e g 25 000 and 12 000 h. Using these tables the normal design calculation procedure described in chapter 18 can be used.
2. For shorter than normal belt life equation 18:5 can be adjusted by an add-on power for belt life $\triangle P_l$ giving:

$$Z=\frac{P_d}{(P_b+\triangle P_{b1}+\triangle P_l)\ k_2\ k_3} \quad \dotfill \quad (19:3)$$

The add-on power for belt life will depend on small pulley diameter d_1 and rpm of small pulley n_1 according to:

$$\triangle P_l=\frac{d_1\ n_1}{k_6} \quad \dotfill \quad (19:4)$$

The coefficient k_6 is determined experimentally for each belt life, belt type and belt section and values for it can be found in power rating tables, where this method of calculation is recommended.

19.7 Belt drives with more complicated geometry involving more than two pulleys

Belt drives containing more than two pulleys will require more complicated calculations than the simple case. As a rule fatigue effects must be calculated for each individual pulley and then summed with application of the cumulative damage theory, cf section 15.7. With the aid of computer programs additional complications can be introduced, e g angular displacement of pulleys, while maintaining the ease of using these programs. These computer programs can also be used in other cases, e g for crossed belt drives or in simple belt drives with idler pulleys.

The starting point for the calculation is a driving shaft with a given rotation speed and power output and a number of driven shafts with given positions, power requirements and rotation speeds. If there are large load variations during operation this is simplest accounted for by a number of load levels with estimated distribution in time. The calculation procedure can then in principle be described as follows:

1. Select proper belt type.
2. Calculate design power by multiplication with the service factor according to equation 18:2.
3. Select diameter of driving pulley as large as belt speed permits. Starting from desired rotation speeds for the driven shafts the corresponding diameters of the driven pulleys are calculated with the aid of equation 17:1. The closest standard pulley diameters are then chosen. If pulleys will be larger than space permits a smaller driving pulley must be chosen and a lower belt speed used. If any of the pulleys end up smaller than the smallest size recommended for the belt section to be used a larger driving pulley may be necessary.
4. With the coordinates for the shafts given and the corresponding pulley diameters known arc of contact angle θ, belt contact length at pulley $\triangle L_\theta$ and belt span length $\triangle L_s$ can now be calculated for all pulleys. The calculations can be performed with the aid of a scale drawing, with equations 11:11 and 11:12 or with a computer program. Belt length may

then be calculated by summing the belt span lengths $\triangle L_s$ and the belt contact lengths at pulleys $\triangle L_\theta$.

5. If shaft coordinates are not fixed but allow a certain variation, corrected coordinates may be chosen, e g to permit space for a large pulley or to fit the standard belt lengths available.

6. Select suitable belt profile. This can be done with the aid of charts, as exemplified in Appendix D.

7. Calculate belt tension force ratio between tight and slack side of the belt $F_{e1}:F_{e2}$ for each pulley with respect to the design power needed at each pulley and its arc of contact angle θ. This can be done with equations of the same type as equation 16:5, but these are usually a part of the computer program used. In these calculations the effective belt tension force in the tight and slack sides of the belt at each pulley will be obtained. By adding these along the belt the tight and slack side belt tension force at the driving pulley will be obtained. If the calculation shows that belt tension ratio at any of the pulleys becomes too large, and thus that there is risk of excessive slip, belt pre-tension has to be increased.

8. Belt life can now be calculated by application of the cumulative damage theory according to the principles outlined in section 15.7. Each pulley is here looked upon with regard to its contribution to the fatigue and further the effect of various fatigue stress levels can be considered and summed. Calculations can either start from a desired belt life and aim at determining the number of belts required to meet this demand or alternatively start from a given number of belts for calculation of the corresponding belt life. For pulleys or idler pulleys used on the outside of the belt a diameter of 75% of the actual one is used in the calculations.

9. Required pre-tension in the belt and shaft loads can be calculated with the aid of computer programs.

The calculation procedure described above is complicated and requires knowledge of a great number of coefficients, why it can practically only be performed with the aid of a computer program. Computer programs are available at most V-belt manufacturers who offer them as a service to their customers.

19.8 Adjustable-speed drives

In an adjustable-speed drive using split, axially adjustable pulleys speed ratio variations of up to 3:1 can be reached with the aid of one split pulley and of up to 9:1 with two split pulleys. Published data on maximum speed ratio variations do vary, however, and considerably lower figures than those mentioned above will often be found. As an adjustable-speed drive usually contains two pulleys operating in the same plane, the same basic design calculation procedure as for other V-belt drives can be used, cf for instance the American standard RMA IP-25. Typical of an adjustable-speed drive is that loading conditions vary depending on the speed ratio variation in the drive. For more accurate calculations fatigue effects of the various loading conditions should therefore be estimated and their effect summed with the aid of the cumulative damage theory, cf section 15.7. In the simplest case three loading levels are chosen: average stress, minimum stress and maximum stress and the corresponding distribution in time estimated. Design calculation can then be performed with the aid of a computer program of the type discussed in section 19.7. In these calculations belt life may be used as a variable. The program is usually made up for a single variator belt but includes the option of various belt profiles. Idler pulleys may also be included. In calculations for adjustable-speed drives the determination of pre-tension is replaced by calculation of the axial force required between the pulley halves, e g calculated according to equation 14:17.

19.9 Resonant frequencies of V-belt vibrations

Design calculations for V-belt drives are usually aimed at determining the data mentioned in this and preceding chapter, i e type and number of belts, diameter of pulleys, standard belt lengths and shaft centre distance with take-up allowances. The calculations can, especially if they are carried out with the aid of computers, be complemented to give other figures of interest. This may include the calculation of resonant frequencies of the belts to avoid belt drive constructions with disturbing resonance phenomena. In resonance the amplitude of vibrations will increase and thus also the noise level.

A belt can vibrate transversally, i e like the string of a guitar or a violin, in which case the resonant frequency f is simplest calculated according to:

$$f = \frac{n_i}{2L} \sqrt{\frac{F}{\varrho A}} \qquad (19:5)$$

The resonant frequency f thus increases with decreasing length of the vibrating string L, with increased tension F in the string and with decreased mass per unit length $A\varrho$ of the string. The factor n_i is an integer, i e $n_i = 1, 2, 3, \ldots$, and indicates that vibrations may appear as multiples of the base frequency. In a belt drive, however, the belt is in movement meaning that equation 19:5 has to be modified. The theories developed for band-saw blades can here be applied. Furthermore it must be observed that belts have a stiffness that cannot be neglected and which takes on different values in the height and width direction of the belt. The resonant frequency will therefore take on different values for these two directions. The equation must also be modified to take care of the special effects that occur when belts leave and enter pulley grooves. The end result will therefore be mathematically rather complicated even when the equation contains a great number of simplifying assumptions.

Belts also show longitudinal vibrations as they form an elastic link between two rotating masses with given moments of inertia on the driving and driven shafts respectively. Equations for the resonant frequency in this case can be set up which contain relevant data, e g the tension stiffness of the belt, its pre-tension and the moment of inertia of the rotating masses.

Transversal vibrations may be initiated by imbalance of belts or pulleys, e g caused by dimensional variations or uneven mass distribution and by improper mounting. Longitudinal vibrations may be initiated by uneven torque loading. To get a belt drive which is as vibration-free as possible these factors have to be counteracted or eliminated. If this is not possible, then the belt drive should be constructed so that resonant frequencies for transversal and longitudinal vibrations do not coincide, otherwise they will tend to intensify each other. Further resonant frequencies corresponding to rotation speeds of belt or pulleys must be avoided as well as belt lengths which are even multiples of pulley circumferences. The amplitude of vibrations will increase with free belt span length and can therefore be reduced by introduction of idler pulleys. Typical values of resonant frequency for transversal vibrations are 20–60 Hz and for longitudinal vibrations 10–20 Hz.

The calculation of resonant frequencies requires fairly complicated equations and large amounts of data for belts, pulleys and the belt drive. Detailed information can be found in design manuals from certain V-belt manufacturers.

Appendix A

List of symbols

(Hz=herz, kg=kilogramme, m=metre, N=newton, Pa=pascal, rad=radian, s=second, W=watt)

Latin characters

Symbol	Quantity	SI unit
A	cross section area	m^2
b_d	datum line differential	m
b_e	effective line differential	m
C_d	datum circumference	m
C_e	effective circumference	m
C_p	pitch circumference	m
c	shaft centre distance	m
c_p	speed ratio	—
c_{ps}	speed ratio with respect to slip and elongation of belt	—
d	diameter	m
d_d	datum diameter (pulley)	m
d_e	effective diameter (pulley)	m
d_p	pitch diameter (pulley)	m
d_1	diameter of small/driving pulley	m
d_2	diameter of large/driven pulley	m
E	(Young's) modulus of elasticity	Pa
e	base of natural (Napierian) logarithms (e=2,718)	—
F	force	N
F_a	axial force	N
F_c	centrifugal force projected along the belt	N
F_{cr}	centrifugal force	N
F_e	effective belt force (effective belt tension)	N

Symbol	Quantity	SI unit
F_m	measuring force	N
F_N	normal force	N
F_V	vertical force	N
F_0	shaft load, bearing load	N
F_1	tight side belt tension	N
F_2	slack side belt tension	N
FR	fatigue rate	s^{-1}
f	resonant frequency	Hz
g	acceleration due to gravity	m/s^2
k (with indices)	constants, coefficients or exponents	
L	belt length	m
L_d	datum length (belt)	m
L_e	effective length (belt)	m
L_i	inner (belt) length	m
L_p	pitch length (belt)	m
L_0	base length (belt)	m
$\triangle L_s$	span length	m
$\triangle L_\theta$	contact length of belt at pulley	m
l	length	m
M	torque	Nm
$\bar{M}$	mean torque	Nm
M_{max}	maximum torque during overload	Nm
M_{s2}	torque on driven shaft with regard to slip	Nm
M_1	torque on driving shaft	Nm
M_2	torque on driven shaft	Nm
m	mass	kg
N	number of deformation cycles	—
N_{ref}	reference number of deformation cycles in the fatigue equation	—
n	rotation speed	s^{-1}
n_1	rotation speed of the driving shaft	s^{-1}
n_2	rotation speed of the driven shaft	s^{-1}
P	power	W
P_b	basic power rating per belt	W
P_d	design power of the drive	W
P_m	prime mover power, motor power, engine power	W
P_{max}	maximum power rating per belt	W
P_1	power transmitted from driving shaft	W
P_2	power uptake at driven shaft	W
$\triangle P_{b1}$	add-on power per belt with respect to speed ratio	W

Symbol	Quantity	SI unit
$\triangle P_{b2}$	add-on power per belt with respect to belt length	W
$\triangle P_l$	add-on power per belt with respect to desired belt life	W
p_b	pitch (synchronous belts and synchronous belt pulleys)	m
q	exponent in the fatigue equation	
r	radius	m
r_{min}	smallest recommended pulley radius	m
r_1	radius of the small/driving pulley	m
r_2	radius of the large/driven pulley	m
$r_{1\,opt}$	radius of small pulley that at given rotation speed gives the largest power rating	m
S	largest value of either speed ratio (c_p) or its inverse value $\left(\dfrac{1}{c_p}\right)$	—
s	slip	—
T	height or thickness of belt section	m
T_i	belt section height below the pitch zone	m
T_y	belt section height above the pitch zone	m
t	time	s
v	belt speed	m/s
w	belt width (maximum), top width	m
w_d	datum width (belt or pulley groove)	m
w_e	effective width (belt or pulley groove)	m
w_p	pitch width (belt or pulley groove)	m
w_w	width of the band of belts in a multiple belt drive	m
Z	number of V-belts in a multiple belt drive	—
z_b	number of teeth in a synchronous belt	—
z_1	number of teeth in the small synchronous belt pulley	—
z_2	number of teeth in the large synchronous belt pulley	—

Greek characters

Symbol	Quantity	SI unit
α	belt angle, pulley groove angle	rad
β	angle during tensiometer measurement of belt tension	rad
δ	deflection during tensiometer measurement of belt tension	m

Symbol	Quantity	SI unit
ε	elongation, strain	—
ε_d	strain difference caused by difference between tight and slack side belt tension	—
ε_i	compressive strain on the underside of the belt due to bending	—
ε_y	elongation strain on the topside of the belt due to bending	—
η	efficiency	—
η_M	efficiency of torque transmission	—
η_{Mb}	load-dependent efficiency of torque transmission	—
η_t	total efficiency	—
η_ω	efficiency of speed transmission	—
θ	arc of contact angle	rad
θ_g	sliding zone of the arc of contact angle	rad
θ_s	static zone of the arc of contact angle	rad
θ_1	arc of contact angle at the small/driving pulley	rad
θ_2	arc of contact angle at the large/driven pulley	rad
λ	effective belt tension ratio	—
λ_b	coefficient of traction	—
μ	coefficient of friction	—
μ_s	apparent coefficient of friction	—
ϱ	density	kg/m^3
σ	stress	Pa
σ_b	bending (flexural) stress	Pa
σ_c	centrifugal stress	Pa
σ_{db}	dynamic stress at break	Pa
σ_{dyn}	dynamic stress	Pa
σ_e	effective tensile stress	Pa
σ_m	average (static) stress	Pa
σ_{ref}	reference stress in the fatigue equation	Pa
σ_t	total stress	Pa
ω	angular velocity	rad/s
ω_1	angular velocity of small/driving pulley	rad/s
ω_2	angular velocity of large/driven pulley	rad/s
$\omega_{1\,opt}$	angular velocity of small pulley that gives maximum power rating at given pulley diameter	rad/s

248

Recommended literature

The literature cited below is recommended for further study. The reference list is obviously not complete. The selection has been made to meet requirements on topicalness, technical standard and that the contents shall offer a deepening or enlargement of the text in this handbook.

Standards

A list of the more important standards in this area follows in Appendix C. Copies of the standards can be ordered from the various national organizations of standardization cited.

Commercial material (catalogues and design manuals)

Most V-belt manufacturers issue catalogues presenting the various types and dimensions of belts produced and give design data and recommended design procedures. More elaborate design manuals are also available, often restricted to certain application areas, e g industrial or agricultural applications. This literature can be obtained from the respective supplier or agent, usually without cost.

Handbooks, textbooks

1. Keilriemen. Eine Monografie. Verlag Ernst Heyer, Essen, 1972, p 1–130.
2. Oberg E, Jones F D, Horton H L: Machinery's Handbook. Industrial Press Inc, New York, 1979 (21st edition), p 1022–1051.
3. Dittrich O, Schumann R: Anwendungen der Antriebstechnik. Band III: Getriebe. Krauss-kopf-Taschenbücher ,,Antriebstechnik", Otto Krausskopf-Verlag, Mainz, 1974, p 139–205, 302–304.

4. The Official Power Transmission Handbook. Published by Power Transmission Distributors Association, Distributor Publications Inc, Stamford, CT, 1981 (second edition), p 41–72.
5. Fronius S: Konstruktionslehre. Antriebselemente. VEB Verlag Technik, Berlin, 1982 (2. Auflage), p 368–395.

Belt types

1. Semin R: The mechanics of belt drives, Part 1. Mach Design, Vol 52, June 1980, p 68–73. Part 2. Mach Design, Vol 52, July 1980, p 77–80.
2. Anon: Belts and belt drives. Mach Design, Vol 53, June 1981, p 28–34.
3. Groothuis I A: Past, present and future of automotive belts. SAE Techn Paper Series 800267, 1980, p 1–10.

Adjustable-speed drives

1. Hitchcox A: Variable speed sheaves – a simple solution to speed control. Power Transm Des, Vol 24, No 1, 1982, p 107–110.
2. Gerbert B G: Adjustable speed V-belt drives – mechanical properties and design. SAE Techn Paper Series 740747, 1974, p 1–10.
3. Gerbert B G: Scheibenspreizkräfte in Breitkeilriemengetrieben. Antriebstechnik *13* (1974):1, p 1–5.

Noise, vibrations

1. Berg P Å: Ljudalstring vid remtransmissioner. Olika remtyper jämförs. Verkstäderna, Vol 73, No 13, Oct 12, 1977, p 27–30. Medd nr 9, 1977, D3.
2. Doyle E, Hornung K G: Lateral vibration of V-belts. ASME publ 69-VIBR-29, 1969, p 1–12.
3. Berg P Å: Noise from different types of belt drives. Inter-Noise 78, San Francisco, May 1978, p 207–212.

Materials and manufacture

1. Donaldson W K: Power transmission belts. Chapter 9 in Wake W C, Wootton D B: Textile reinforcement of elastomers. Applied Science Publishers Ltd, London, 1982, p 225–259.
2. Rogers J W: The use of fibers in V-belt compounds. Rubb World, Vol 183, No 6, March 1981, p 27–31.
3. Mirza J: Polychloroprene in power transmission belting. American Chem Soc, Rubb Div 125th Meeting, Indianapolis, IN, May 1984, Paper No 71, p 1–25.
4. Stanhope H W: V-belt reinforcement – polyester, the most popular fiber. American Chem Soc, Rubb Div 125th Meeting, Indianapolis, IN, May 1984, Paper No 82, p 1–41.

5. Henderson J L, Campbell J H: Polyester elastomer in power transmission belts – a review of current technology. American Chem Soc, Rubb Div 125th Meeting, Indianapolis, IN, May 1984, Paper No 81, p 1–16.

Standardization

1. Erickson W: New standards for power transmission belts. Mach Design, Vol 53, Jan 1981, p 83–89.

Forces and deformations in belts and between belts and pulleys

1. Gerbert B G: Some notes on V-belt drives. ASME publ 80-C2/DET-91, 1980, p 1–11.
2. Gerbert B G: Pressure distribution and belt deformation in V-belt drives. Trans ASME. J Engng for Ind, Vol 97, Aug 1975, p 976–982.
3. Gerbert B G: Tensile stress distribution in the cord of V-belts. Trans ASME. J Engng for Ind, Vol 97, Series B, No 1, Feb 1975, p 14–22.
4. Gerbert B G: Force and slip behaviour in V-belt drives. Acta Polytechnica Scandinavica, Mech Engng Series No 67, Helsink, 1972, p 1–101.

Belt tension and pre-tension

1. Fuetterer W G, Schlamadinger J: Belt tension decay of automotive V- and V-ribbed belts. SAE Techn Paper Series 800851, 1980, p 1–15.
2. Nawrocki W: V-belt tensioning. Mach Design, Vol 43, May 1971, p 72–76.
3. Orito T: Genesis of the maintenance-free belt. SAE Techn Paper Series 820490, 1982, p 1–8.

Fatigue and belt life

1. Shepherd J D, Jackson L L: V-belt reliability – a statistical study of large sample size fatigue tests. SAE Techn Paper Series 800446, 1980, p 1–7.
2. Loecke H: Beitrag zur Ermittlung der übertragbaren Leistung von Keilriemen. Kaut, Gummi, Kunstst 30 (1977):7, p 27–32.
3. Oliver L R, Johnson C O, Breig W F: V-belt life prediction and power rating. Trans ASME. J Engng for Ind, Vol 98, Series B, Feb 1976, p 340–347.
4. Marco S M, Starkey W L, Hornung K G: A quantitative investigation of the factors which influence the fatigue life of a V-belt. Trans ASME. J Engng for Ind, Vol 82, Series B, No 1, Feb 1960, p 47–59.

Friction, slip and power transmission

1. Webber J S: The effective friction coefficient of belts. J IRI, Vol 6, Dec 1972, p 245–248.
2. Belofsky H: On the theory of power transmission by V-belts. Wear *39* (1976) p 263–275.
3. Gerbert B G: A note on slip in V-belt drives. Trans ASME. J Engng for Ind, Vol 98, Series B, No 4, Nov 1976, p 1366–1368.
4. Gerbert B G: A complementary large slip solution in V-belt mechanics. ASME publ. 77-DET-162, 1977, p 1–8.
5. Oliver L R, Breig W F, Hall D D: Calculating V-belt torque capacity. Power Transm Des, Vol 20, No 4, 1978, p 39–41.
6. Erickson W: Cutting drive size and cost with narrow V-belts. Mach Design, Vol 51, No 2, Jan 1979, p 116–119.

Efficiency, energy losses

1. Gilbert K T, Cohen M, Houser D R: Energy transfer in loaded V-belt systems. ASME publ 80-C2/DET-99, 1980, p 1–13.
2. Semin R: Efficiency of belt drives. Mach Design, Vol 53, April 1981, p 197–201.
3. Gerbert B G: Power loss and optimum tensioning of V-belt drives. Trans ASME. J Engng for Ind, Vol 96, Series B, No 3, Aug 1974, p 877–885.
4. Breig W F, Oliver L R: Energy loss and efficiency of power transmission belts. Proc National Conf Power Transmission, Annual Meeting, Chicago, IL, 1980, Vol 7, p 173–183.

Design calculations

1. Pronin B A, Ovchinnikova V A: Calculation of vee-belt drives. Soviet Eng Res, Vol 2, No 3, 1982, p 19–22.
2. Motherway J E: Designing V-belt drives with a microcomputer. Comput Mech Eng, Vol 2, No 1, 1983, p 20–27.

Important standards for pulleys and belts

Below important standards are first listed in numerical order for each organization of standardization. A table is also presented from which relevant standards can be identified for certain application areas, belt types and standardized properties. All standards mentioned are in the English language with the following exceptions: The Swedish SIS/SMS standards are only in Swedish with exception for those having ISO numbers which utilize the English ISO text. German DIN standards are primarily in the German language but English titles are available and have been introduced into the list. For some of the German standards complete English translations are available which is indicated by "(En)" after the title. Translations of DIN standards into other languages are also available. It should further be mentioned that all ISO standards are available also in French versions.

International Organization for Standardization (ISO)

ISO	22-1975	Flat transmission belts and corresponding pulleys – Widths
ISO	63-1975	Flat transmission belts – Lengths
ISO	99-1975	Pulleys for flat transmission belts – Diameters
ISO	100-1975	Pulleys for flat transmission belts – Crowns
ISO	155-1975	Transmission pulleys – Limiting values for adjustment of centres
ISO	254-1981	Quality, finish and balance of transmission pulleys
ISO	255-1981	Pulleys for classical and narrow V-belts – Geometrical inspection of grooves
ISO	1081-1980	Drives using V-belts and grooved pulleys – Terminology
ISO	1604-1976	Endless wide V-belts for industrial speed-changers and groove profiles for corresponding pulleys
ISO	1813-1979	Antistatic endless V-belts – Electrical conductivity – Characteristic and method of test
ISO	2790-1982	Narrow V-belt drives for the automotive industry – Dimensions
ISO	3410-1976	Agricultural machinery – Endless variable-speed V-belts and groove sections of corresponding pulleys

ISO 4183-1980   Grooved pulleys for classical and narrow V-belts
ISO 4184-1980   Classical and narrow V-belts – Lengths
ISO 5287-1978   Narrow V-belt drives for the automotive industry – Fatigue test
ISO 5288-1982   Synchronous belt drives – Vocabulary
ISO 5289-1978   Endless hexagonal belts for agricultural machinery, and groove sections of
    corresponding pulleys
ISO 5290-1978   Grooved pulleys for joined narrow V-belts – Groove sections 9J, 15J, 20J and
    25J
ISO 5292-1980   Industrial V-belt drives – Calculation of power ratings
ISO 5294-1979   Synchronous belt drives – Pulleys
ISO 5295-1981   Synchronous belts – Calculation of power rating and drive centre distance
ISO 5296-1978   Synchronous belt drives – Belts
    Addendum 1-1982
    Addendum 2-1982

Swedish national standard:
Sveriges Mekanstandardisering (SMS)/
Standardiseringskommissionen i Sverige (SIS)

The title of each standard is preceded by "Remmar och remskivor", after which follows the text
given below.

SMS 979-1962   Ändlösa flata transmissionsremmar och rembanor. Dimensioner
SMS 2475-1967   Remskivor. Utförande. Balansering
SMS 2476-1967   Beräkning av kilremsväxel. Profil Y, Z, A, B, C, D, E
SMS 2477-1970   Kilremmar. Profil Y, Z, A, B, C, D, E
SMS 2479-1967   Kilremmar profil Y. Effektöverföring
SMS 2480-1967   Kilremmar profil Z. Effektöverföring
SMS 2481-1967   Kilremmar profil A. Effektöverföring
SMS 2482-1967   Kilremmar profil B. Effektöverföring
SMS 2483-1967   Kilremmar profil C. Effektöverföring
SMS 2484-1967   Kilremmar profil D. Effektöverföring
SMS 2485-1967   Kilremmar profil E. Effektöverföring
SMS 2489-1967   Kontroll av kilremsprofiler. Profil Y, Z, A, B, C, D, E
SMS 2490-1967   Kontroll av kilremslängd. Profil Y, Z, A, B, C, D, E
SMS 2491-1967   Kilremsskivor. Bearbetade kilremsspår. Profil Y, Z, A, B, C, D, E
SMS 2492-1967   Kilremsskivor. Delningsdiametrar och utväxlingar. Profil Y, Z, A, B, C, D,
    E
SMS 2493-1967   Bearbetade kilremsskivor. Kontroll av spår
SMS 2494-1973   Kilremsskivor. Ritningsdata
SMS 2500-1975   Kilremmar och kilremsskivor. Terminologi
SMS 2516-1975   Smalkilremmar. Profil SPZ, SPB, SPC
SMS 2517-1975   Smalkilremmar. Profil SPZ, SPB, SPC. Kontroll av remlängd
SMS 2518-1975   Kilremsskivor. Profil SPZ, SPB, SPC. Bearbetade kilremsspår

SMS 2565-1972   Kilremmar. Profil W16-W100
SMS 2566-1972   Kilremmar. Profil W16-W100. Kontroll av remlängd och profil
SMS 2567-1972   Kilremsskivor. Profil W16-W100. Bearbetade kilremsspår
SS-ISO 3410-1981 Remmar och remskivor för lantbruksmaskiner – Ändlösa hastighetsvariabla
                 kilremmar samt spår i kilremsskivor (Eng)
SS-ISO 5289-1981 Remmar och remskivor för lantbruksmaskiner – Ändlösa sexkantiga remmar
                 samt spår i remskivor (Eng)

Rubber Manufacturers Association (RMA)/ Mechanical Power Transmission Association (MPTA)/The Rubber Association of Canada (RAC)

Engineering Standards

Specifications for drives using:

IP-20-1977   Classical multiple V-belts (A, B, C, D, and E cross sections)
IP-21-1984   Double-V (hexagonal) belts (AA, BB, CC, DD cross sections)
IP-22-1983   Narrow multiple V-belts (3V, 5V, and 8V cross sections)
IP-23-1968   Single V-belts (2L, 3L, 4L, and 5L cross sections)
IP-24-1983   Synchronous belts (MXL, XL, L, H, XH, and XXH belt sections)
IP-25-1982   Variable speed belts (12 cross sections)
IP-26-1977   V-ribbed belts (H, J, K, L, and M cross sections)

German national standard: DIN Deutsches Institut für Normung

"(En)" indicates that a complete English version is available.

DIN 109 Teil 1-1973   Antriebselemente. Umfangsgeschwindigkeiten (En)
                      Driving elements. Circumferential speeds
DIN 109 Teil 2-1973   Antriebselemente. Achsabstände für Riementriebe mit Keilriemen
                      (En)
                      Driving elements. Centre distances for V-belt drives
DIN 111-1982   Antriebselemente. Flachriemenscheiben. Masse, Nenndrehmomente
                      (En)
                      Driving elements. Pulleys for flat transmission belts.
                      Dimensions, nominal torsional moments

DIN 2211 Teil 1-1982   Antriebselemente. Schmalkeilriemenscheiben. Masse, Werkstoff
Driving components. Grooved pulleys for narrow V-belts. Dimensions, materials

DIN 2211 Teil 2-1973   Antriebselemente. Schmalkeilriemenscheiben. Prüfung der Rillen
Driving components. Grooved pulleys for narrow V-belts. Verification

DIN 2211 Teil 3-1974   Antriebselemente. Schmalkeilriemenscheiben. Zuordnung für elektrische Maschinen (En)
Driving elements. Grooved pulleys for narrow V-belts. Correlation to electric motors

DIN 2215-1975   Endlose Keilriemen. Masse (En)
Endless V-belts. Dimensions

DIN 2216-1972   Endliche Keilriemen. Masse (En)
Open ended V-belts. Dimensions

DIN 2217 Teil 1-1973   Antriebselemente. Keilriemenscheiben. Masse, Werkstoff (En)
Driving elements. V-belt pulleys. Dimensions, material

DIN 2217 Teil 2-1973   Antriebselemente. Keilriemenscheiben. Prüfung der Rillen (En)
Driving elements. Grooved pulleys for V-belts. Testing of grooves

DIN 2218-1976   Endlose Keilriemen für den Maschinenbau. Berechnung der Antriebe, Leistungswerte (En)
Endless V-belts for mechanical engineering. Calculation of drives, power ratings

DIN 7719 Teil 1-1983   Endlose Breitkeilriemen für industrielle Drehzahlwandler. Riemen und Rillenprofile der zugehörigen Scheiben
Endless wide V-belts for industrial speed changers.
V-belts and groove profiles for corresponding pulleys

DIN 7721 Teil 1-1979   Synchronriementriebe, metrische Teilung. Synchronriemen
Synchronous belt drives, metric pitch. Synchronous belts

DIN 7721 Teil 2-1979   Synchronriementriebe, metrische Teilung. Zahnlückenprofil für Synchronscheiben
Synchronous belt drives, metric pitch. Tooth space profile of synchronous pulleys

DIN 7722-1982   Endlose Hexagonalriemen für Landmaschinen und Rillenprofile der zugehörigen Scheiben
Endless hexagonal belts for agricultural machinery and groove sections of corresponding pulleys

DIN 7753 Teil 1-1977   Endlose Schmalkeilriemen für den Maschinenbau. Masse (En)
Endless narrow V-belts for industrial purposes. Dimensions

DIN 7753 Teil 2-1976   Endlose Schmalkeilriemen für den Maschinenbau. Berechnung der Antriebe, Leistungswerte
Narrow V-belts for industrial purposes. Calculation of drives, power ratings

DIN 7753 Teil 3-1983   Endlose Schmalkeilriemen für den Kraftfahrzeugbau. Riemen und Scheibenrillenprofile
Endless narrow V-belts for the automotive industry
Belts and pulley groove profiles

DIN 7753 Teil 4-1978   Endlose Schmalkeilriemen für den Kraftfahrzeugbau. Ermüdungsprüfung
Endless narrow V-belts for automotive drives. Fatigue test

DIN-ISO 5290-1981	Rillenscheiben für Verbund-Schmalkeilriemen. Rillenprofile 9J, 15J, 20J und 25J (En)
	Grooved pulleys for joined narrow V-belts. Groove sections 9J, 15J, 20J and 25J
DIN-ISO 5294-1983	Synchronriementriebe. Scheiben (En)
	Synchronous belt drives. Pulleys
DIN-ISO 5296-1983	Synchronriementriebe. Riemen (En)
	Synchronous belt drives. Belts

British Standards Institution (BSI)

BS 351-1976	Specification for rubber, balata or plastics flat transmission belting of textile construction for general use
BS 3733-1974	Specification for endless V-belt drives for agricultural purposes
BS 3790-1981	Specification for endless wedge belt drives and endless V-belt drives
BS 4548-1970	Specification for synchronous belt drives
BS AU 150-1973	Automobile series
	Specification for automotive V-belt drives

Society of Automotive Engineers (SAE)

SAE J636c-1977	V-belts and pulleys
SAE J637b-1974	Automotive V-belt drives
SAE J720-1980	Tractor belt speed and pulley width
SAE J1278-1980	SI (metric) synchronous belts and pulleys
SAE J1313-1980	Automotive synchronous belt drives

American Society of Agricultural Engineers (ASAE)

| ASAE S211.3-1978 | V-belt drives for agricultural machines |
| ASAE S212.1-1978 | Laboratory procedure for testing V-belts |

CONTENTS OF IMPORTANT STANDARDS FOR PULLEYS AND BELTS WITH REFERENCE TO BELT TYPE, APPLICATION AREA AND STANDARDIZED PROPERTIES

Belt type	Application area	Belts						Pulleys					Design calculation of belt drives
		Section dimensions	Measurement of section dimensions	Length	Length measurement	Power rating (T=tabulated data, B=calculation formulae)	Other data	Dimensions of grooves, teeth or crown	Measurement of grooves, teeth or crown	Diameter, width or hub dimensions	Measurement of diameter, width or hub	Workmanship and finish	
Classical V-belt	Industrial	RMA-IP-20 DIN 2215 DIN 2216	SMS 2489 RMA-IP-20	ISO 4184 SMS 2477 RMA-IP-20 DIN 2215 BS 3790	ISO 4184 SMS 2490 RMA-IP-20 DIN 2215 BS 3790	ISO 5292 (B) SMS 2479-85 (T) RMA-IP-20 (T, B) DIN 2218 (T) BS 3790 (T)	ISO 1813 (el cond) BS 3790 (el cond, flammability)	ISO 4183 SMS 2491 RMA-IP-20 DIN 2217 BS 3790	ISO 255 SMS 2493 RMA-IP-20 DIN 2217	ISO 4183 SMS 2491 SMS 2492 RMA-IP-20 DIN 2217 BS 3790	ISO 255 SMS 2493 RMA-IP-20 DIN 2217	ISO 254 SMS 2475 SMS 2494 RMA-IP-20 DIN 2217 BS 3790	SMS 2476 RMA-IP-20 DIN 2218 BS 3790
	Agricultural	BS 3733 ASAE S211.3		BS 3733 ASAE S211.3	BS 3733 ASAE S211.3		ASAE S212.1 (fatigue test)	BS 3733 ASAE S211.3		BS 3733		BS 3733 ASAE S211.3	BS 3733
	Home appliances	RMA-IP-23	RMA-IP-23	RMA-IP-23	RMA-IP-23	RMA-IP-23 (T, B)		RMA-IP-23	RMA-IP-23	RMA-IP-23	RMA-IP-23	RMA-IP-23	
Narrow V-belt	Industrial	SMS 2516 RMA-IP-22 DIN 7753 Teil 1–2	RMA-IP-22	ISO 4184 SMS 2516 RMA-IP-22 DIN 7753 Teil 1–2 BS 3790	ISO 4184 SMS 2517 RMA-IP-22 DIN 7753 Teil 1–2 BS 3790	ISO 5292 (B) RMA-IP-22 (T, B) DIN 7753 (T) Teil 1–2 BS 3790 (T)	BS 3790 (el cond, flammability)	ISO 4183 SMS 2518 RMA-IP-22 DIN 2211 BS 3790	ISO 255 SMS 2493 RMA-IP-22 DIN 2211	ISO 4183 SMS 2518 RMA-IP-22 DIN 2211 BS 3790	ISO 255 SMS 2493 RMA-IP-22 DIN 2211	ISO 254 SMS 2475 SMS 2494 RMA-IP-22 DIN 2211 BS 3790	RMA-IP-22 DIN 7753 Teil 1–2 BS 3790
	Automotive	ISO 2790 DIN 7753 Teil 3–4 BS AU 150	BS AU 150 SAE J636c	DIN 7753 Teil 3–4 BS AU 150 SAE J636c	ISO 2790 DIN 7753 Teil 3–4 BS AU 150 SAE J636c		ISO 5287 (fatigue test) DIN 7753 Teil 3–4 (fatigue test) BS AU 150 (fatigue test) SAE J637b (fatigue test)	ISO 2790 DIN 2211 BS AU 150 SAE J636c	DIN 2211 BS AU 150	DIN 2211 BS AU 150 SAE J636c	ISO 2790 DIN 2211 BS AU 150	DIN 2211 BS AU 150	BS AU 150

Belt type	Application area	Belts						Pulleys					Design calculation of belt drives
		Section dimensions	Measurement of section dimensions	Length	Length measurement	Power rating (T=tabulated data. B= calculation formulae)	Other data	Dimensions of grooves, teeth or crown	Measurement of grooves, teeth or crown	Diameter, width or hub dimensions	Measurement of diameter, width or hub	Workmanship and finish	
Wide V-belt	Industrial	ISO 1604 SMS 2565 RMA-IP-25 DIN 7719 Teil 1	ISO 1604 SMS 2566 RMA-IP-25 DIN 7719 Teil 1	ISO 1604 SMS 2565 RMA-IP-25 DIN 7719 Teil 1	ISO 1604 SMS 2566 RMA-IP-25 DIN 7719 Teil 1	RMA-IP-25 (T. B)		ISO 1604 SMS 2567 RMA-IP-25 DIN 7719 Teil 1	ISO 1604 RMA-IP-25	ISO 1604 SMS 2567 RMA-IP-25 DIN 7719 Teil 1	ISO 1604	SMS 2567 RMA-IP-25	RMA-IP-25
	Agricultural	ISO 3410 BS 3733 ASAE S211.3	ISO 3410	ISO 3410 BS 3733 ASAE S211.3	BS 3733 ASAE S 211.3		ASAE S212.1 (fatigue test)	ISO 3410 BS 3733 ASAE S211.3		ISO 3410 BS 3733		BS 3733 ASAE S211.3	
Hexagonal belt	Industrial	RMA-IP-21	RMA-IP-21	RMA-IP-21	RMA-IP-21			RMA-IP-21	RMA-IP-21	RMA-IP-21	RMA-IP-21	ISO 254 RMA-IP-21	
	Agricultural	ISO 5289 DIN 7722 BS 3733 ASAE S211.3		ISO 5289 DIN 7722 BS 3733 ASAE S211.3	ISO 5289 DIN 7722 BS 3733 ASAE S211.3		ASAE S212.1 (fatigue test)	ISO 5289 DIN 7722 BS 3733 ASAE S211.3		ISO 5289 BS3733		BS 3733 ASAE S211.3	BS 3733
Joined V-belt – Classical profile	Industrial	RMA-IP-20	RMA-IP-20	RMA-IP-20	RMA-IP-20			RMA-IP-20	RMA-IP-20	RMA-IP-20	RMA-IP-20	RMA-IP-20	
– Narrow profile	Industrial	RMA-IP-22	RMA-IP-22	RMA-IP-22	RMA-IP-22			ISO 5290 RMA-IP-22	RMA-IP-22	ISO 5290 RMA-IP-22	RMA-IP-22	ISO 254 RMA-IP-22	

Belt type	Application area	Belts						Pulleys					Design calculation of belt drives
		Section dimensions	Measurement of section dimensions	Length	Length measurement	Power rating (T=tabulated data. B=calculation formulae)	Other data	Dimensions of grooves. teeth or crown	Measurement of grooves. teeth or crown	Diameter. width or hub dimensions	Measurement of diameter. width or hub	Workmanship and finish	
Synchronous belt	Industrial	ISO 5296 RMA-IP-24 DIN 7721 BS 4548		ISO 5296 RMA-IP-24 DIN 7721 BS 4548	ISO 5296 RMA-IP-24 DIN 7721 BS 4548	ISO 5296 (B) RMA-IP-24 (T. B) BS 4548 (T. B)	ISO 5296 DAD 2 (doublesided)	ISO 5294 RMA-IP-24 DIN 7721 BS 4548		ISO 5294 RMA-IP-24 DIN 7721 BS 4548		ISO 254 RMA-IP-24 BS 4548	RMA-IP-24 BS 4548
	Automotive	SAE J1278 SAE J1313			SAE J1278			SAE J1278 SAE J1313	SAE J1313	SAE J1278 SAE J1313		SAE J1313	
Flat belt	Industrial	ISO 22 SMS 979 BS 351		ISO 63 SMS 979	BS 351	BS 351 (T. B)		ISO 22 ISO 100 SMS 979 DIN 111		ISO 99 SMS 979 DIN 111 BS 351		ISO 254 DIN 111	BS 351
V-ribbed belt	General	RMA-IP-26		RMA-IP-26	RMA-IP-26	RMA-IP-26 (T. B)		RMA-IP-26	RMA-IP-26	RMA-IP-26	RMA-IP-26	RMA-IP-26	RMA-IP-26

Selection chart for V-belt profiles

Below a few examples are given of charts for the selection of suitable belt profile starting from the rotation speed of the small pulley n_1 (rpm) and the design power P_d (kW). The example is for cut V-belts (TrellPower X from Trelleborg AB). The addition of the letter "X" at the end of the profile designation is here used to indicate cut V-belts. In the chart of narrow V-belts the profile designations 3VX and 5VX from American standard RMA IP-22 are also used. The design calculation example from chapter 18 with n_1=4400 rpm and P_d=36 kW is marked.

CHART 1
Narrow V-belts

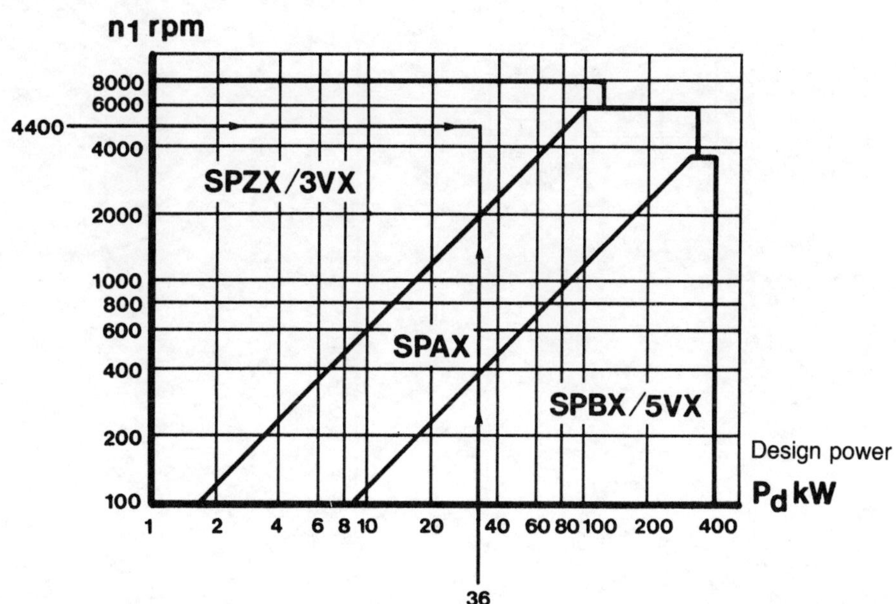

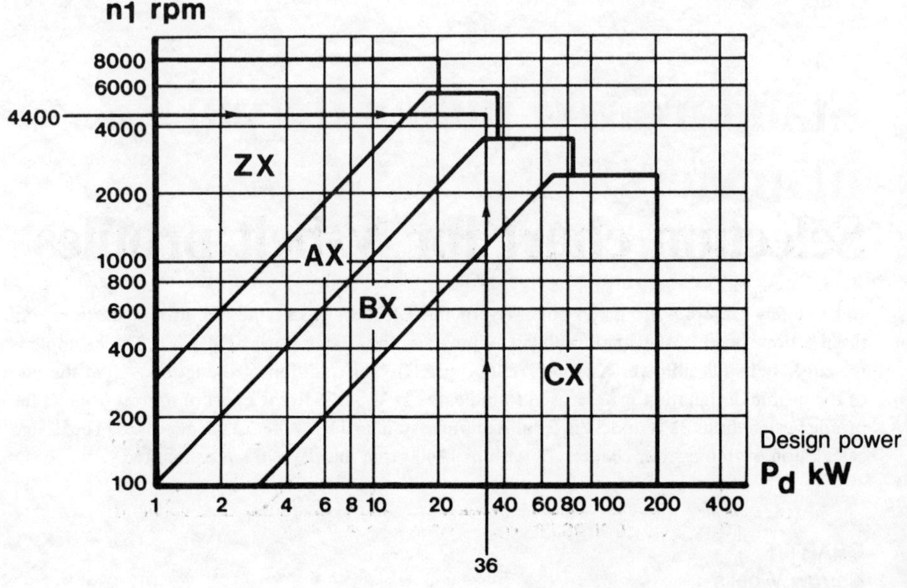

n1 rpm
8000
6000
4400
4000
2000
1000
800
600
400
200
100
ZX
AX
BX
CX
Design power
Pd kW
1
2
4
6
8 10
20
40
60 80 100
200
400
36

Standardized pulley datum diameters

Below standardized datum diameters for pulleys up to 500 mm diameter according to ISO 4183 are given. On the nominal dimensions a tolerance of $\pm 0,8\%$ is accepted. In practical design calculations it is usually best to take data from manufacturers' catalogues which may also contain data on number of grooves, material and construction, hub design for connection to shaft and availability (stock item or made to order).

The symbols used indicate:
+ only recommended for classical V-belts
* can be used for classical as well as for narrow V-belts
without marking: not recommended

Datum diameters Nominal values mm	Degree of preference for datum diameters according to the groove profiles						
	Y	Z SPZ	A SPZ	B SPB	C SPC	D	E
20	+						
22,4	+						
25	+						
28	+						
31,5	+						
35,5	+						
40	+						
45	+						
50	+	+					
53							
56	+	+					
60							
63	+	*					
67							
71	+	*					
75		*	+				
80	+	*	+				
85			+				
90	+	*	*				
95			*				
100	+	*	*				
106			*				
112	+	*	*				
118			*				
125	+	*	*	+			

Datum diameters Nominal values mm	Degree of preference for datum diameters according to the groove profiles						
	Y	Z SPZ	A SPZ	B SPB	C SPC	D	E
132		*	*	+			
140		*	*	*			
150		*	*	*			
160		*	*	*			
170				*			
180		*	*	*			
190							
200		*	*	*	+		
212					+		
224		*	*	*	*		
236					*		
250		*	*	*	*		
265					*		
280		*	*	*	*		
300					*		
315		*	*	*	*		
335					*		
355		*	*	*	*	+	
375						+	
400		*	*	*	*	+	
425						+	
450		*	*	*	*	+	
475						+	
500		*	*	*	*	+	+

Standard belt lengths

The examples shown below are for common profiles of cut V-belts (data for TrellPower X from Trelleborg AB). The complete catalogue also gives inside and outside belt lengths and in addition contains data for a number of profiles which are without interest for the design calculation example presented in chapter 18. ISO 4184 gives the standard length as the datum length L_d which for most practical purposes is equivalent to the pitch length L_p. The table also shows that manufacturers can offer a number of belt lengths which cannot be traced to national or international standards but were once developed and made to order for special applications.

SPZX/3VX

Profile designation and belt marking	Pitch length L_p, mm	Datum length L_d according to ISO 4184, mm	Profile designation and belt marking	Pitch length L_p, mm	Datum length L_d according to ISO 4184, mm
SPZX 612	612	—	SPZX 1362	1362	—
SPZX 630	630	630	SPZX 1387	1387	—
SPZX 650	650	—	SPZX 1400	1400	1400
SPZX 670	670	—	SPZX 1415	1415	—
SPZX 687	687	—	SPZX 1437	1437	—
SPZX 710	710	710	SPZX 1462	1462	—
SPZX 737	737	—	SPZX 1500	1500	—
SPZX 750	750	—	SPZX 1520	1520	—
SPZX 762	762	—	SPZX 1537	1537	—
SPZX 787	787	—	SPZX 1562	1562	—
SPZX 800	800	800	SPZX 1600	1600	1600
SPZX 825	825	—	SPZX 1637	1637	—
SPZX 837	837	—	SPZX 1662	1662	—
SPZX 850	850	—	SPZX 1700	1700	—
SPZX 862	862	—	SPZX 1737	1737	—
SPZX 887	887	—	SPZX 1762	1762	—
SPZX 900	900	900	SPZX 1800	1800	1800
SPZX 912	912	—	SPZX 1837	1837	—
SPZX 937	937	—	SPZX 1862	1862	—
SPZX 950	950	—	SPZX 1900	1900	—
SPZX 962	962	—	SPZX 1937	1937	—
SPZX 987	987	—	SPZX 1962	1962	—
SPZX 1000	1000	1000	SPZX 2000	2000	2000
SPZX 1010	1010	—	SPZX 2025	2025	—
SPZX 1037	1037	—	SPZX 2062	2062	—
SPZX 1060	1060	—	SPZX 2120	2120	—
SPZX 1075	1075	—	SPZX 2150	2150	—
SPZX 1087	1087	—	SPZX 2240	2240	2240
SPZX 1112	1112	—	SPZX 2280	2280	—
SPZX 1120	1120	1120	SPZX 2360	2360	—
SPZX 1140	1140	—	SPZX 2400	2400	—
SPZX 1162	1162	—	SPZX 2500	2500	2500
SPZX 1180	1180	—	SPZX 2535	2535	—
SPZX 1200	1200	—	SPZX 2650	2650	—
SPZX 1212	1212	—	SPZX 2685	2685	—
SPZX 1237	1237	—	SPZX 2800	2800	2800
SPZX 1250	1250	1250	SPZX 2840	2840	—
SPZX 1265	1265	—	SPZX 3000	3000	—
SPZX 1287	1287	—	SPZX 3150	3150	3150
SPZX 1320	1320	—	SPZX 3350	3350	—
SPZX 1340	1340	—	SPZ —	—	3550

SPAX

Profile designation and belt marking	Pitch length L_p, mm	Datum length L_d according to ISO 4184, mm	Profile designation and belt marking	Pitch length L_p, mm	Datum length L_d according to ISO 4184, mm
SPAX 732	732	—	SPAX 1700	1700	—
SPAX 800	800	800	SPAX 1732	1732	—
SPAX 832	832	—	SPAX 1757	1757	—
SPAX 850	850	—	SPAX 1800	1800	1800
SPAX 882	882	—	SPAX 1832	1832	—
SPAX 900	900	900	SPAX 1857	1857	—
SPAX 950	950	—	SPAX 1900	1900	—
SPAX 982	982	—	SPAX 1932	1932	—
SPAX 1000	1000	1000	SPAX 1957	1957	—
SPAX 1032	1032	—	SPAX 2000	2000	2000
SPAX 1060	1060	—	SPAX 2032	2032	—
SPAX 1082	1082	—	SPAX 2057	2057	—
SPAX 1107	1107	—	SPAX 2120	2120	—
SPAX 1120	1120	1120	SPAX 2182	2182	—
SPAX 1132	1132	—	SPAX 2240	2240	2240
SPAX 1157	1157	—	SPAX 2300	2300	—
SPAX 1180	1180	—	SPAX 2360	2360	—
SPAX 1207	1207	—	SPAX 2432	2432	—
SPAX 1232	1232	—	SPAX 2482	2482	—
SPAX 1250	1250	1250	SPAX 2500	2500	2500
SPAX 1282	1282	—	SPAX 2532	2532	—
SPAX 1320	1320	—	SPAX 2607	2607	—
SPAX 1357	1357	—	SPAX 2650	2650	—
SPAX 1382	1382	—	SPAX 2682	2682	—
SPAX 1400	1400	1400	SPAX 2800	2800	2800
SPAX 1432	1432	—	SPAX 2882	2882	—
SPAX 1457	1457	—	SPAX 3000	3000	—
SPAX 1500	1500	—	SPAX 3150	3150	3150
SPAX 1532	1532	—	SPAX 3350	3350	—
SPAX 1557	1557	—	SPA —	—	3550
SPAX 1600	1600	1600	SPA —	—	4000
SPAX 1632	1632	—	SPA —	—	4500
SPAX 1657	1657	—			

AX

Profile designation and belt marking	Pitch length L_p, mm	Datum length L_d according to ISO 4184, mm	Profile designation and belt marking	Pitch length L_p, mm	Datum length L_d according to ISO 4184, mm
AX 22	604	—	AX 56	1454	—
AX 23	622	—	AX 57	1497	—
A —	—	630	AX 58	1512	—
AX 24	642	—	AX 59	1529	—
AX 25	662	—	A —	—	1550
AX 26	679	—	AX 60	1554	—
A —	—	700	AX 61	1597	—
AX 27	702	—	AX 62	1604	—
AX 28	742	—	AX 63	1629	—
AX 29	754	—	A —	—	1640
AX 30	779	—	AX 64	1654	—
A —	—	790	AX 65	1679	—
AX 31	804	—	AX 66	1697	—
AX 32	842	—	AX 67	1729	—
AX 33	879	—	A —	—	1750
A —	—	890	AX 68	1754	—
AX 34	892	—	AX 69	1779	—
AX 36	942	—	AX 70	1804	—
A —	—	990	AX 71	1829	—
AX 38	992	—	AX 72	1854	—
AX 39	1029	—	AX 73	1879	—
AX·40	1052	—	AX 74	1909	—
AX 41	1967	—	AX 75	1929	—
A —	—	1100	A —	—	1940
AX 42	1104	—	AX 76	1954	—
AX 43	1132	—	AX 77	1979	—
AX 44	1154	—	AX 78	2017	—
AX 45	1172	—	AX 79	2029	—
AX 46	1192	—	A —	—	2050
AX 47	1229	—	AX 80	2054	—
AX 48	1247	—	AX 81	2079	—
A —	—	1250	AX 82	2117	—
AX 49	1279	—	AX 83	2147	—
AX 50	1317	—	AX 84	2164	—
AX 51	1332	—	AX 85	2179	—
AX 52	1354	—	A —	—	2200
AX 53	1379	—	AX 86	2209	—
AX 54	1407	—	AX 87	2237	—
AX 55	1429	—	AX 88	2272	—
A —	—	1430	AX 89	2297	—

AX

Profile designation and belt marking	Pitch length L_p, mm	Datum length L_d according to ISO 4184, mm	Profile designation and belt marking	Pitch length L_p, mm	Datum length L_d according to ISO 4184, mm
A —	—	2300	AX 104	2679	—
AX 90	2309	—	A —	—	2700
AX 91	2329	—	AX 105	2704	—
AX 92	2357	—	AX 108	2779	—
AX 93	2392	—	AX 110	2832	—
AX 94	2429	—	AX 112	2879	—
AX 95	2459	—	AX 118	3029	—
A —	—	2480	AX 120	3079	—
AX 97	2497	—	AX 125	3195	—
AX 98	2529	—	AX 128	3279	—
AX 100	2569	—	AX 132	3379	—
AX 101	2604	—			

BX

Profile designation and belt marking	Pitch length L_p, mm	Datum length L_d according to ISO 4184, mm	Profile designation and belt marking	Pitch length L_p, mm	Datum length L_d according to ISO 4184, mm
BX 24	657	—	BX 63	1649	—
BX 26	697	—	BX 64	1674	—
BX 27	724	—	BX 65	1692	—
BX 28	749	—	BX 66	1724	—
BX 29	774	—	BX 67	1749	—
BX 30	799	—	B —	—	1760
BX 31	824	—	BX 68	1774	—
BX 32	849	—	BX 69	1792	—
BX 33	874	—	BX 70	1824	—
B —	—	930	BX 71	1849	—
BX 36	949	—	BX 72	1874	—
BX 37	974	—	BX 73	1892	—
BX 38	997	—	BX 74	1924	—
B —	—	1000	BX 75	1949	—
BX 39	1024	—	B —	—	1950
BX 40	1062	—	BX 76	1974	—
BX 41	1074	—	BX 77	1992	—
B —	—	1100	BX 78	2024	—
BX 42	1107	—	BX 79	2049	—
BX 43	1127	—	BX 80	2074	—
BX 44	1167	—	BX 81	2099	—
BX 45	1187	—	BX 82	2112	—
BX 46	1199	—	BX 83	2142	—
B —	—	1210	BX 84	2174	—
BX 47	1242	—	B —	—	2180
BX 48	1252	—	BX 85	2204	—
BX 49	1274	—	BX 86	2232	—
BX 50	1312	—	BX 87	2267	—
BX 51	1327	—	BX 88	2292	—
B —	—	1370	B —	—	2300
BX 52	1374	—	BX 89	2304	—
BX 53	1392	—	BX 90	2324	—
BX 54	1424	—	BX 91	2352	—
BX 55	1449	—	BX 92	2387	—
BX 57	1492	—	BX 94	2424	—
BX 58	1524	—	BX 95	2454	—
BX 59	1549	—	BX 96	2474	—
B —	—	1560	B —	—	2500
BX 61	1592	—	BX 97	2504	—
BX 62	1624	—	BX 98	2524	—

BX

Profile designation and belt marking	Pitch length L_p, mm	Datum length L_d according to ISO 4184, mm	Profile designation and belt marking	Pitch length L_p, mm	Datum length L_d according to ISO 4184, mm
BX 101	2599	—	BX 117	3024	—
BX 102	2642	—	BX 119	3074	—
BX 103	2656	—	BX 124	3190	—
BX 104	2674	—	B —	—	3200
BX 105	2699	—	BX 127	3274	—
B —	—	2700	BX 131	3374	—
BX 107	2774	—	B —	—	3600
BX 108	2792	—	B —	—	4060
BX 110	2827	—	B —	—	4430
B —	—	2870	B —	—	4820
BX 112	2874	—	B —	—	5370
BX 114	2929	—	B —	—	6070
BX 116	2992	—			

Basic power rating per belt

Below some examples are given of basic power rating per belt P_b (kW) and add-on power for speed ratio $\triangle P_{b1}$ (kW) for some V-belt profiles as a function of small pulley diameter d_{pl} (mm) and rotation speed n_1 (rpm). The thick, stepped lines in the chart correspond to (from the upper left corner) belt speeds v of 10, 20, 30 and 40 m/s. The example is for cut V-belts (TrellPower X from Trelleborg AB) and for a belt life of 25000 h. The complete catalogue also contains data for other profiles, which are not of interest for the design calculation example given in chapter 18, and data for belt life of 12000 h.

Basic power rating P_b per belt (kW) at different pitch diameters d_{p1} (mm) of the small pulley · Add-on power $\triangle P_{b1}$ (kW) for varying speed ratio

Rotation speed of small pulley n_1 (rpm)	56	63	67	71	75	80	85	90	95	100	106	112	118	125	132	140	150	160	180	1,00 -1,01	1,02 -1,03	1,04 -1,06	1,07 -1,09	1,10 -1,13	1,14 -1,18	1,19 -1,25	1,26 -1,35	1,36 -1,57	≥ 1,58
200	0,17	0,22	0,25	0,28	0,30	0,34	0,37	0,40	0,44	0,47	0,51	0,55	0,59	0,63	0,68	0,73	0,79	0,85	0,98	0,00	0,00	0,00	0,01	0,01	0,01	0,01	0,02	0,02	0,02
400	0,32	0,41	0,46	0,51	0,56	0,63	0,69	0,75	0,82	0,88	0,95	1,03	1,10	1,19	1,28	1,37	1,50	1,62	1,86	0,00	0,00	0,01	0,01	0,02	0,02	0,03	0,03	0,04	0,04
600	0,45	0,58	0,66	0,73	0,81	0,90	0,99	1,09	1,18	1,27	1,38	1,49	1,60	1,72	1,85	1,99	2,17	2,34	2,70	0,00	0,01	0,01	0,02	0,03	0,03	0,04	0,05	0,06	0,06
800	0,57	0,75	0,84	0,94	1,04	1,16	1,28	1,40	1,52	1,64	1,79	1,93	2,07	2,23	2,40	2,58	2,81	3,04	3,50	0,00	0,01	0,02	0,03	0,04	0,05	0,06	0,06	0,08	0,08
1000	0,69	0,90	1,02	1,15	1,27	1,41	1,56	1,71	1,86	2,01	2,18	2,36	2,53	2,73	2,93	3,16	3,44	3,72	4,28	0,00	0,01	0,02	0,04	0,05	0,06	0,07	0,08	0,09	0,11
1200	0,80	1,06	1,20	1,34	1,48	1,66	1,84	2,01	2,19	2,36	2,57	2,77	2,98	3,21	3,45	3,72	4,05	4,38	5,03	0,00	0,01	0,03	0,04	0,06	0,07	0,08	0,10	0,11	0,13
1400	0,91	1,20	1,37	1,53	1,70	1,90	2,10	2,30	2,50	2,70	2,94	3,18	3,41	3,68	3,96	4,26	4,64	5,02	5,76	0,00	0,02	0,03	0,05	0,06	0,08	0,10	0,11	0,13	0,15
1600	1,02	1,34	1,53	1,72	1,90	2,13	2,36	2,59	2,81	3,04	3,31	3,57	3,84	4,14	4,45	4,79	5,22	5,64	6,47	0,00	0,02	0,04	0,06	0,07	0,09	0,11	0,13	0,15	0,17
1800	1,12	1,48	1,69	1,90	2,10	2,36	2,61	2,87	3,12	3,37	3,66	3,96	4,25	4,59	4,93	5,31	5,78	6,24	7,15	0,00	0,02	0,04	0,06	0,08	0,10	0,13	0,15	0,17	0,19
2000	1,21	1,62	1,85	2,08	2,30	2,58	2,86	3,14	3,41	3,69	4,01	4,34	4,66	5,03	5,39	5,81	6,32	6,82	7,81	0,00	0,02	0,05	0,07	0,09	0,12	0,14	0,16	0,19	0,21
2200	1,31	1,75	2,00	2,25	2,49	2,80	3,10	3,41	3,70	4,00	4,35	4,70	5,05	5,45	5,85	6,30	6,85	7,39	8,44	0,00	0,03	0,05	0,08	0,10	0,13	0,15	0,18	0,21	0,23
2400	1,40	1,88	2,15	2,42	2,68	3,01	3,34	3,67	3,99	4,31	4,69	5,06	5,44	5,86	6,29	6,77	7,35	7,93	9,04	0,00	0,03	0,06	0,09	0,11	0,14	0,17	0,20	0,23	0,25
2600	1,49	2,00	2,29	2,58	2,87	3,22	3,57	3,92	4,26	4,61	5,01	5,41	5,81	6,27	6,72	7,22	7,84	8,45	9,61	0,00	0,03	0,06	0,09	0,12	0,15	0,18	0,21	0,24	0,28
2800	1,58	2,13	2,44	2,74	3,05	3,42	3,80	4,17	4,53	4,90	5,33	5,75	6,17	6,65	7,13	7,66	8,31	8,94	10,15	0,00	0,03	0,07	0,10	0,13	0,16	0,20	0,23	0,26	0,30
3000	1,67	2,25	2,57	2,90	3,22	3,62	4,02	4,41	4,80	5,18	5,63	6,08	6,52	7,03	7,53	8,08	8,76	9,42	10,66	0,00	0,04	0,07	0,11	0,14	0,17	0,21	0,25	0,28	0,32
3200	1,75	2,36	2,71	3,05	3,39	3,82	4,23	4,65	5,05	5,46	5,93	6,40	6,86	7,39	7,91	8,49	9,19	9,86	11,14	0,00	0,04	0,08	0,12	0,15	0,19	0,22	0,26	0,30	0,34
3400	1,83	2,48	2,84	3,20	3,56	4,01	4,44	4,88	5,30	5,72	6,22	6,71	7,19	7,74	8,28	8,87	9,59	10,29	11,58	0,00	0,04	0,08	0,12	0,16	0,20	0,24	0,28	0,32	0,36
3600	1,91	2,59	2,97	3,35	3,73	4,19	4,65	5,10	5,55	5,98	6,50	7,01	7,51	8,08	8,63	9,24	9,98	10,68	11,98	0,00	0,04	0,09	0,13	0,17	0,21	0,25	0,30	0,34	0,38
3800	1,98	2,69	3,10	3,49	3,88	4,37	4,85	5,32	5,78	6,24	6,77	7,30	7,81	8,40	8,96	9,59	10,34	11,05	12,34	0,00	0,05	0,09	0,14	0,17	0,22	0,27	0,31	0,36	0,40
4000	2,06	2,80	3,22	3,63	4,04	4,54	5,04	5,53	6,01	6,48	7,03	7,57	8,10	8,70	9,28	9,92	10,68	11,39	12,67	0,00	0,05	0,10	0,14	0,18	0,23	0,28	0,33	0,38	0,42
4200	2,13	2,90	3,34	3,77	4,19	4,71	5,23	5,73	6,23	6,71	7,28	7,84	8,38	8,99	9,58	10,22	10,99	11,69	12,95	0,00	0,05	0,10	0,15	0,19	0,24	0,29	0,34	0,39	0,45
4400	2,20	3,00	3,45	3,90	4,34	4,88	5,41	5,93	6,44	6,94	7,52	8,09	8,64	9,27	9,86	10,51	11,27	11,97	13,18	0,00	0,05	0,11	0,16	0,20	0,26	0,31	0,36	0,41	0,47
4600	2,26	3,10	3,56	4,02	4,48	5,04	5,58	6,12	6,64	7,16	7,75	8,33	8,89	9,52	10,12	10,77	11,53	12,21	13,37	0,00	0,06	0,11	0,17	0,21	0,27	0,32	0,38	0,43	0,49
4800	2,33	3,19	3,67	4,15	4,62	5,19	5,75	6,30	6,84	7,36	7,97	8,56	9,13	9,76	10,36	11,01	11,75	12,42		0,00	0,06	0,12	0,17	0,22	0,28	0,34	0,39	0,45	0,51
5000	2,39	3,28	3,78	4,27	4,75	5,34	5,92	6,48	7,03	7,56	8,18	8,78	9,35	9,99	10,59	11,23	11,95	12,60		0,00	0,06	0,12	0,18	0,23	0,29	0,35	0,41	0,47	0,53
5200	2,45	3,36	3,88	4,38	4,88	5,48	6,07	6,65	7,21	7,75	8,38	8,98	9,55	10,19	10,79	11,42	12,12	12,73		0,00	0,06	0,12	0,19	0,24	0,30	0,36	0,43	0,49	0,55
5400	2,50	3,45	3,98	4,49	5,00	5,62	6,22	6,81	7,38	7,93	8,56	9,17	9,74	10,38	10,96	11,58	12,26			0,00	0,06	0,13	0,19	0,25	0,31	0,38	0,44	0,51	0,57
5600	2,56	3,53	4,07	4,60	5,12	5,75	6,36	6,96	7,54	8,09	8,73	9,34	9,91	10,54	11,12	11,72	12,36			0,00	0,07	0,13	0,20	0,26	0,32	0,39	0,46	0,53	0,59
5800	2,61	3,61	4,16	4,70	5,23	5,88	6,50	7,11	7,69	8,25	8,89	9,50	10,07	10,69	11,25	11,83				0,00	0,07	0,14	0,21	0,27	0,34	0,41	0,48	0,55	0,61
6000	2,66	3,68	4,25	4,80	5,34	5,99	6,63	7,24	7,83	8,39	9,04	9,64	10,21	10,81	11,36	11,91				0,00	0,07	0,14	0,22	0,28	0,35	0,42	0,49	0,56	0,64
6200	2,71	3,75	4,33	4,89	5,44	6,11	6,75	7,37	7,96	8,53	9,17	9,77	10,33	10,92	11,44					0,00	0,07	0,15	0,22	0,29	0,36	0,43	0,51	0,58	0,66
6400	2,75	3,82	4,41	4,98	5,54	6,22	6,87	7,49	8,08	8,65	9,29	9,88	10,43	11,00	11,50					0,00	0,08	0,15	0,23	0,29	0,37	0,45	0,52	0,60	0,68
6600	2,79	3,88	4,48	5,06	5,63	6,32	6,97	7,60	8,20	8,76	9,39	9,98	10,51	11,06						0,00	0,08	0,16	0,24	0,30	0,38	0,46	0,54	0,62	0,70
6800	2,83	3,94	4,55	5,14	5,72	6,41	7,07	7,70	8,30	8,86	9,49	10,06	10,57	11,10						0,00	0,08	0,16	0,24	0,31	0,39	0,48	0,56	0,64	0,72
7000	2,87	4,00	4,62	5,22	5,80	6,50	7,16	7,79	8,39	8,94	9,56	10,12	10,62							0,00	0,08	0,17	0,25	0,32	0,41	0,49	0,57	0,66	0,74
7200	2,90	4,05	4,68	5,29	5,87	6,58	7,24	7,87	8,47	9,02	9,62	10,16	10,64							0,00	0,09	0,17	0,26	0,33	0,42	0,50	0,59	0,68	0,76
7400	2,94	4,10	4,74	5,35	5,94	6,65	7,32	7,95	8,53	9,08	9,67	10,19								0,00	0,09	0,18	0,27	0,34	0,43	0,52	0,61	0,70	0,78
7600	2,96	4,14	4,79	5,41	6,01	6,72	7,38	8,01	8,59	9,12	9,70	10,20								0,00	0,09	0,18	0,27	0,35	0,44	0,53	0,62	0,71	0,81
7800	2,99	4,19	4,84	5,46	6,06	6,77	7,44	8,06	8,64	9,16	9,71									0,00	0,09	0,19	0,28	0,36	0,45	0,55	0,64	0,73	0,83
8000	3,01	4,22	4,88	5,51	6,11	6,83	7,49	8,11	8,67	9,18	9,71									0,00	0,10	0,19	0,29	0,37	0,46	0,56	0,66	0,75	0,85

Basic power rating P_b per belt (kW) at different pitch diameters d_{p1} (mm) of the small pulley — and **Add-on power $\triangle P_{b1}$ (kW) for varying speed ratio**

Rotation speed of small pulley n_1 (rpm)	71	75	80	85	90	95	100	106	112	118	125	132	140	150	160	180	200	224	250	1,00 -1,01	1,02 -1,03	1,04 -1,06	1,07 -1,09	1,10 -1,13	1,14 -1,18	1,19 -1,25	1,26 -1,35	1,36 -1,57	≥ 1,58
200	0,27	0,32	0,37	0,43	0,48	0,54	0,60	0,66	0,73	0,79	0,87	0,95	1,03	1,14	1,25	1,46	1,68	1,93	2,20	0,00	0,01	0,01	0,02	0,02	0,03	0,03	0,04	0,04	0,05
400	0,48	0,56	0,67	0,78	0,89	0,99	1,10	1,22	1,35	1,48	1,62	1,77	1,94	2,14	2,35	2,76	3,16	3,65	4,16	0,00	0,01	0,02	0,03	0,04	0,06	0,07	0,08	0,09	0,10
600	0,66	0,78	0,94	1,10	1,25	1,41	1,56	1,75	1,93	2,12	2,33	2,54	2,79	3,09	3,39	3,98	4,57	5,27	6,02	0,00	0,02	0,03	0,05	0,07	0,08	0,10	0,12	0,13	0,15
800	0,82	0,99	1,19	1,40	1,60	1,80	2,01	2,25	2,49	2,73	3,01	3,28	3,60	3,99	4,38	5,16	5,92	6,83	7,80	0,00	0,02	0,04	0,07	0,09	0,11	0,13	0,15	0,18	0,20
1000	0,97	1,17	1,43	1,68	1,93	2,18	2,43	2,73	3,02	3,32	3,66	4,00	4,38	4,86	5,34	6,29	7,22	8,32	9,50	0,00	0,03	0,06	0,08	0,11	0,14	0,17	0,19	0,22	0,25
1200	1,11	1,35	1,65	1,95	2,24	2,54	2,83	3,19	3,53	3,88	4,29	4,69	5,14	5,71	6,27	7,38	8,47	9,76	11,13	0,00	0,03	0,07	0,10	0,13	0,17	0,20	0,23	0,27	0,30
1400	1,24	1,52	1,86	2,21	2,55	2,89	3,23	3,63	4,03	4,43	4,89	5,35	5,88	6,52	7,17	8,43	9,68	11,14	12,69	0,00	0,04	0,08	0,11	0,15	0,19	0,23	0,27	0,31	0,35
1600	1,36	1,68	2,07	2,45	2,84	3,22	3,61	4,06	4,51	4,96	5,48	6,00	6,59	7,31	8,03	9,45	10,84	12,46	14,16	0,00	0,04	0,09	0,13	0,18	0,22	0,27	0,31	0,35	0,40
1800	1,48	1,83	2,26	2,69	3,12	3,55	3,97	4,48	4,98	5,48	6,05	6,63	7,28	8,08	8,87	10,43	11,95	13,71	15,55	0,00	0,05	0,10	0,15	0,20	0,25	0,30	0,35	0,40	0,45
2000	1,59	1,97	2,45	2,92	3,39	3,86	4,33	4,88	5,43	5,98	6,61	7,23	7,94	8,82	9,68	11,37	13,00	14,89	16,85	0,00	0,06	0,11	0,16	0,22	0,28	0,33	0,39	0,44	0,50
2200	1,69	2,11	2,63	3,14	3,65	4,16	4,67	5,27	5,87	6,46	7,14	7,82	8,58	9,53	10,45	12,26	14,01	16,01	18,06	0,00	0,06	0,12	0,18	0,24	0,30	0,37	0,42	0,49	0,55
2400	1,78	2,24	2,80	3,35	3,91	4,46	5,00	5,65	6,29	6,92	7,66	8,38	9,20	10,21	11,20	13,12	14,95	17,05	19,16	0,00	0,07	0,13	0,20	0,26	0,33	0,40	0,46	0,53	0,60
2600	1,87	2,36	2,96	3,56	4,15	4,74	5,32	6,01	6,70	7,37	8,16	8,93	9,80	10,86	11,91	13,92	15,84	18,00	20,15	0,00	0,07	0,15	0,21	0,29	0,36	0,43	0,50	0,57	0,65
2800	1,96	2,47	3,12	3,75	4,38	5,01	5,63	6,36	7,09	7,81	8,63	9,45	10,36	11,49	12,58	14,68	16,67	18,87	21,03	0,00	0,08	0,16	0,23	0,31	0,39	0,46	0,54	0,62	0,70
3000	2,03	2,58	3,26	3,94	4,61	5,27	5,92	6,70	7,47	8,22	9,09	9,95	10,91	12,08	13,22	15,39	17,42	19,65	21,79	0,00	0,08	0,17	0,25	0,33	0,41	0,50	0,58	0,66	0,75
3200	2,10	2,68	3,40	4,12	4,82	5,52	6,21	7,02	7,83	8,62	9,53	10,42	11,42	12,64	13,82	16,05	18,11	20,34	22,42	0,00	0,09	0,18	0,26	0,35	0,44	0,53	0,62	0,71	0,80
3400	2,17	2,78	3,54	4,29	5,02	5,76	6,48	7,33	8,17	9,00	9,95	10,87	11,91	13,16	14,38	16,65	18,73	20,92	22,91	0,00	0,10	0,19	0,28	0,37	0,47	0,56	0,66	0,75	0,85
3600	2,23	2,87	3,66	4,45	5,22	5,98	6,74	7,63	8,50	9,36	10,34	11,30	12,37	13,66	14,89	17,20	19,27	21,40		0,00	0,10	0,20	0,30	0,40	0,50	0,60	0,69	0,80	0,90
3800	2,28	2,95	3,78	4,60	5,40	6,20	6,98	7,91	8,81	9,70	10,71	11,70	12,80	14,11	15,37	17,69	19,73	21,77		0,00	0,11	0,21	0,31	0,42	0,52	0,63	0,73	0,84	0,95
4000	2,33	3,03	3,89	4,74	5,58	6,40	7,21	8,17	9,11	10,02	11,06	12,08	13,19	14,53	15,80	18,11	20,10			0,00	0,11	0,22	0,33	0,44	0,55	0,66	0,77	0,88	1,00
4200	2,37	3,10	3,99	4,87	5,74	6,60	7,43	8,42	9,38	10,32	11,39	12,42	13,56	14,91	16,18	18,47	20,39			0,00	0,12	0,24	0,34	0,46	0,58	0,70	0,81	0,93	1,05
4400	2,41	3,16	4,09	5,00	5,90	6,78	7,64	8,65	9,64	10,60	11,69	12,74	13,89	15,25	16,51	18,76				0,00	0,12	0,25	0,36	0,48	0,61	0,73	0,85	0,97	1,10
4600	2,44	3,21	4,17	5,11	6,04	6,94	7,83	8,87	9,88	10,86	11,97	13,03	14,18	15,54	16,80	18,98				0,00	0,13	0,26	0,38	0,51	0,63	0,76	0,89	1,02	1,15
4800	2,46	3,26	4,25	5,22	6,17	7,10	8,01	9,07	10,10	11,10	12,21	13,28	14,44	15,80	17,03					0,00	0,13	0,27	0,39	0,53	0,66	0,80	0,93	1,06	1,20
5000	2,48	3,30	4,32	5,32	6,29	7,24	8,17	9,25	10,30	11,31	12,44	13,51	14,67	16,00	17,21					0,00	0,14	0,28	0,41	0,55	0,69	0,83	0,97	1,10	1,24
5200	2,49	3,34	4,38	5,40	6,40	7,37	8,32	9,42	10,48	11,50	12,63	13,70	14,85	16,16	17,33					0,00	0,15	0,29	0,43	0,57	0,72	0,86	1,00	1,15	1,29
5400	2,49	3,36	4,43	5,48	6,50	7,49	8,45	9,56	10,63	11,66	12,80	13,86	15,00	16,28						0,00	0,15	0,30	0,44	0,59	0,75	0,90	1,04	1,19	1,34
5600	2,49	3,38	4,48	5,54	6,58	7,59	8,56	9,69	10,77	11,80	12,93	13,99	15,10	16,34						0,00	0,16	0,31	0,46	0,62	0,77	0,93	1,08	1,24	1,39
5800	2,48	3,39	4,51	5,60	6,65	7,67	8,66	9,80	10,88	11,91	13,04	14,08	15,16							0,00	0,16	0,32	0,48	0,64	0,80	0,96	1,12	1,28	1,44

Rotation speed of small pulley n_1 (rpm)	Basic power rating P_b per belt (kW) at different pitch diameters d_{p1} (mm) of the small pulley																	Add-on power ΔP_{b1} (kW) for varying speed ratio									
	75	80	85	90	95	100	106	112	118	125	132	140	150	160	180	200	224	1,00 -1,02	1,03 -1,07	1,08 -1,13	1,14 -1,21	1,22 -1,30	1,31 -1,44	1,45 -1,64	1,65 -2,01	2,02 -3,00	≥3,01
200	0,37	0,41	0,45	0,49	0,52	0,56	0,61	0,65	0,70	0,75	0,80	0,86	0,93	1,00	1,14	1,28	1,44	0,00	0,01	0,02	0,02	0,03	0,04	0,05	0,06	0,07	0,08
400	0,63	0,70	0,77	0,84	0,91	0,98	1,06	1,14	1,22	1,32	1,41	1,51	1,64	1,77	2,02	2,26	2,55	0,00	0,02	0,03	0,05	0,07	0,08	0,10	0,12	0,13	0,15
600	0,85	0,95	1,05	1,15	1,25	1,34	1,46	1,57	1,68	1,81	1,94	2,09	2,27	2,45	2,79	3,13	3,53	0,00	0,03	0,05	0,07	0,10	0,13	0,15	0,18	0,20	0,23
800	1,05	1,18	1,30	1,43	1,55	1,67	1,82	1,96	2,10	2,27	2,43	2,61	2,84	3,06	3,50	3,92	4,42	0,00	0,03	0,07	0,10	0,13	0,17	0,20	0,23	0,27	0,30
1000	1,22	1,37	1,53	1,68	1,82	1,97	2,14	2,32	2,49	2,68	2,87	3,09	3,36	3,63	4,14	4,64	5,22	0,00	0,04	0,08	0,12	0,17	0,21	0,25	0,29	0,33	0,38
1200	1,38	1,56	1,73	1,90	2,08	2,25	2,45	2,64	2,84	3,07	3,29	3,54	3,85	4,15	4,74	5,30	5,96	0,00	0,05	0,10	0,15	0,20	0,25	0,30	0,35	0,40	0,45
1400	1,52	1,72	1,92	2,12	2,31	2,50	2,73	2,95	3,17	3,42	3,67	3,95	4,30	4,63	5,28	5,91	6,62	0,00	0,06	0,12	0,17	0,23	0,29	0,35	0,41	0,47	0,53
1600	1,65	1,88	2,10	2,31	2,53	2,74	2,99	3,23	3,48	3,76	4,03	4,34	4,71	5,08	5,78	6,45	7,21	0,00	0,07	0,13	0,20	0,27	0,33	0,40	0,47	0,53	0,60
1800	1,77	2,02	2,26	2,49	2,73	2,96	3,23	3,50	3,76	4,06	4,36	4,69	5,09	5,49	6,24	6,94	7,73	0,00	0,08	0,15	0,22	0,30	0,38	0,45	0,53	0,60	0,68
2000	1,88	2,15	2,41	2,66	2,91	3,16	3,46	3,74	4,03	4,35	4,66	5,02	5,45	5,86	6,65	7,37	8,17	0,00	0,08	0,17	0,25	0,33	0,42	0,50	0,58	0,67	0,75
2200	1,98	2,26	2,54	2,82	3,09	3,35	3,66	3,97	4,27	4,61	4,95	5,32	5,77	6,20	7,01	7,75	8,53	0,00	0,09	0,18	0,27	0,37	0,46	0,55	0,64	0,73	0,83
2400	2,07	2,37	2,67	2,96	3,25	3,53	3,86	4,18	4,49	4,85	5,20	5,59	6,05	6,50	7,32	8,06	8,82	0,00	0,10	0,20	0,30	0,40	0,50	0,60	0,70	0,80	0,90
2600	2,15	2,47	2,78	3,09	3,39	3,69	4,03	4,37	4,70	5,07	5,43	5,83	6,31	6,76	7,58	8,30	9,02	0,00	0,11	0,22	0,32	0,43	0,54	0,65	0,76	0,87	0,98
2800	2,22	2,56	2,89	3,21	3,52	3,83	4,19	4,54	4,88	5,27	5,64	6,05	6,53	6,98	7,80	8,49	9,13	0,00	0,12	0,23	0,35	0,47	0,59	0,70	0,82	0,94	1,05
3000	2,29	2,64	2,98	3,32	3,64	3,96	4,34	4,70	5,05	5,44	5,82	6,23	6,72	7,17	7,96	8,60	9,15	0,00	0,13	0,25	0,37	0,50	0,63	0,75	0,88	1,00	1,13
3200	2,34	2,71	3,06	3,41	3,75	4,08	4,47	4,84	5,19	5,60	5,98	6,39	6,87	7,31	8,07	8,64	9,08	0,00	0,13	0,27	0,40	0,53	0,67	0,80	0,93	1,07	1,20
3400	2,39	2,77	3,14	3,50	3,85	4,19	4,58	4,96	5,32	5,72	6,11	6,52	6,99	7,41	8,12	8,61	8,90	0,00	0,14	0,28	0,42	0,57	0,71	0,85	0,99	1,14	1,28
3600	2,43	2,82	3,20	3,57	3,93	4,27	4,67	5,06	5,42	5,83	6,21	6,61	7,07	7,47	8,11	8,50	8,62	0,00	0,15	0,30	0,45	0,60	0,75	0,90	1,05	1,20	1,35
3800	2,46	2,86	3,25	3,63	4,00	4,35	4,75	5,14	5,51	5,91	6,28	6,68	7,11	7,49	8,04	8,31	8,23	0,00	0,16	0,32	0,47	0,63	0,79	0,95	1,11	1,27	1,43
4000	2,48	2,90	3,30	3,68	4,05	4,41	4,82	5,20	5,57	5,97	6,33	6,71	7,12	7,46	7,91	8,03		0,00	0,17	0,33	0,50	0,67	0,84	1,00	1,17	1,34	1,50
4200	2,50	2,92	3,33	3,72	4,10	4,46	4,86	5,25	5,61	6,00	6,35	6,71	7,08	7,38	7,71	7,67		0,00	0,18	0,35	0,52	0,70	0,88	1,05	1,23	1,40	1,58
4400	2,51	2,94	3,35	3,75	4,13	4,49	4,89	5,27	5,63	6,00	6,34	6,67	7,00	7,25	7,44			0,00	0,18	0,37	0,55	0,73	0,92	1,10	1,28	1,47	1,65
4600	2,51	2,94	3,36	3,76	4,14	4,50	4,91	5,28	5,62	5,98	6,30	6,60	6,88	7,07	7,11			0,00	0,19	0,38	0,57	0,77	0,96	1,15	1,34	1,54	1,73
4800	2,50	2,94	3,36	3,76	4,14	4,50	4,90	5,26	5,59	5,93	6,22	6,49	6,72	6,84				0,00	0,20	0,40	0,60	0,80	1,00	1,20	1,40	1,60	1,80
5000	2,48	2,93	3,35	3,75	4,13	4,48	4,87	5,23	5,54	5,86	6,12	6,34	6,51	6,55				0,00	0,21	0,42	0,62	0,83	1,04	1,25	1,46	1,67	1,88
5200	2,46	2,91	3,33	3,73	4,11	4,45	4,83	5,17	5,46	5,75	5,98	6,16	6,25	6,21				0,00	0,22	0,43	0,65	0,87	1,09	1,30	1,52	1,74	1,96
5400	2,42	2,88	3,30	3,70	4,07	4,40	4,77	5,09	5,36	5,62	5,80	5,93	5,95					0,00	0,23	0,45	0,67	0,90	1,13	1,35	1,58	1,80	2,03
5600	2,38	2,83	3,26	3,65	4,01	4,34	4,69	4,99	5,23	5,45	5,60	5,66	5,60					0,00	0,23	0,47	0,70	0,94	1,17	1,40	1,64	1,87	2,11
5800	2,33	2,78	3,20	3,59	3,94	4,25	4,58	4,86	5,08	5,26	5,35	5,36						0,00	0,24	0,48	0,72	0,97	1,21	1,45	1,69	1,94	2,18

Rotation speed of small pulley n_1 (rpm)	Basic power rating P_b per belt (kW) at different pitch diameters d_{p1} (mm) of the small pulley																	Add-on power ΔP_{b1} (kW) for varying speed ratio									
	118	125	132	140	150	160	170	180	190	200	212	224	250	280	315	355	400	1,00 -1,02	1,03 -1,07	1,08 -1,13	1,14 -1,21	1,22 -1,30	1,31 -1,44	1,45 -1,64	1,65 -2,01	2,02 -3,00	≥3,01
100	0,53	0,57	0,61	0,65	0,71	0,76	0,82	0,87	0,93	0,98	1,04	1,11	1,24	1,40	1,57	1,77	1,99	0,00	0,01	0,01	0,02	0,03	0,03	0,04	0,04	0,05	0,06
200	0,94	1,01	1,09	1,17	1,27	1,37	1,47	1,57	1,67	1,76	1,88	1,99	2,24	2,52	2,84	3,19	3,59	0,00	0,01	0,03	0,04	0,05	0,06	0,08	0,09	0,10	0,12
300	1,31	1,41	1,52	1,63	1,78	1,92	2,06	2,20	2,34	2,47	2,64	2,80	3,14	3,53	3,98	4,48	5,03	0,00	0,02	0,04	0,06	0,08	0,10	0,12	0,13	0,15	0,17
400	1,65	1,78	1,92	2,06	2,25	2,43	2,61	2,78	2,96	3,13	3,34	3,54	3,98	4,47	5,04	5,66	6,35	0,00	0,03	0,05	0,08	0,10	0,13	0,15	0,18	0,21	0,23
500	1,97	2,13	2,29	2,47	2,69	2,90	3,12	3,33	3,54	3,75	4,00	4,25	4,77	5,36	6,03	6,77	7,58	0,00	0,03	0,06	0,10	0,13	0,16	0,19	0,22	0,26	0,29
600	2,27	2,46	2,64	2,85	3,11	3,36	3,61	3,86	4,10	4,34	4,63	4,91	5,52	6,20	6,96	7,81	8,73	0,00	0,04	0,08	0,12	0,15	0,19	0,23	0,27	0,31	0,35
700	2,56	2,77	2,98	3,22	3,51	3,79	4,08	4,35	4,63	4,90	5,23	5,55	6,23	6,99	7,85	8,79	9,80	0,00	0,04	0,09	0,13	0,18	0,22	0,27	0,31	0,36	0,40
800	2,84	3,07	3,30	3,56	3,89	4,21	4,52	4,83	5,14	5,44	5,80	6,15	6,90	7,74	8,68	9,70	10,80	0,00	0,05	0,10	0,15	0,20	0,26	0,31	0,36	0,41	0,46
900	3,10	3,36	3,61	3,90	4,25	4,60	4,95	5,29	5,62	5,95	6,35	6,73	7,54	8,45	9,46	10,55	11,71	0,00	0,06	0,12	0,17	0,23	0,29	0,35	0,40	0,46	0,52
1000	3,35	3,63	3,91	4,22	4,61	4,99	5,36	5,73	6,09	6,45	6,87	7,28	8,16	9,12	10,19	11,34	12,54	0,00	0,06	0,13	0,19	0,26	0,32	0,38	0,45	0,51	0,58
1100	3,60	3,90	4,20	4,53	4,95	5,35	5,75	6,15	6,54	6,92	7,37	7,81	8,74	9,76	10,88	12,07	13,29	0,00	0,07	0,14	0,21	0,28	0,35	0,42	0,49	0,56	0,63
1200	3,83	4,15	4,47	4,83	5,27	5,71	6,13	6,55	6,96	7,37	7,84	8,31	9,29	10,35	11,51	12,73	13,96	0,00	0,08	0,15	0,23	0,31	0,39	0,46	0,54	0,62	0,69
1300	4,05	4,40	4,74	5,12	5,59	6,05	6,50	6,94	7,37	7,80	8,30	8,79	9,80	10,91	12,10	13,33	14,53	0,00	0,08	0,17	0,25	0,33	0,42	0,50	0,58	0,67	0,75
1400	4,27	4,63	4,99	5,39	5,89	6,37	6,85	7,31	7,77	8,21	8,73	9,24	10,29	11,43	12,63	13,85	15,02	0,00	0,09	0,18	0,27	0,36	0,45	0,54	0,63	0,72	0,81
1500	4,48	4,86	5,24	5,66	6,18	6,69	7,18	7,67	8,14	8,60	9,14	9,67	10,75	11,90	13,11	14,31	15,41	0,00	0,10	0,19	0,29	0,38	0,48	0,58	0,67	0,77	0,87
1600	4,68	5,08	5,48	5,92	6,46	6,99	7,50	8,01	8,50	8,98	9,53	10,07	11,18	12,34	13,54	14,69	15,70	0,00	0,10	0,20	0,31	0,41	0,51	0,61	0,72	0,82	0,92
1700	4,88	5,29	5,70	6,16	6,73	7,28	7,81	8,33	8,84	9,33	9,90	10,45	11,57	12,73	13,91	15,00	15,88	0,00	0,11	0,22	0,33	0,44	0,55	0,65	0,76	0,87	0,98
1800	5,06	5,50	5,92	6,40	6,98	7,55	8,10	8,64	9,16	9,66	10,24	10,80	11,93	13,09	14,22	15,23	15,96	0,00	0,12	0,23	0,35	0,46	0,58	0,69	0,81	0,92	1,04
1900	5,24	5,69	6,13	6,63	7,23	7,81	8,38	8,93	9,46	9,97	10,56	11,13	12,26	13,39	14,48	15,38	15,92	0,00	0,12	0,24	0,36	0,49	0,61	0,73	0,85	0,97	1,10
2000	5,41	5,88	6,34	6,85	7,47	8,07	8,65	9,21	9,75	10,27	10,86	11,43	12,55	13,65	14,67	15,44	15,76	0,00	0,13	0,26	0,38	0,51	0,64	0,77	0,90	1,03	1,15
2100	5,58	6,06	6,53	7,05	7,69	8,30	8,90	9,47	10,01	10,54	11,14	11,70	12,81	13,87	14,79	15,41	15,49	0,00	0,13	0,27	0,40	0,54	0,67	0,81	0,94	1,08	1,21
2200	5,74	6,23	6,72	7,25	7,90	8,53	9,13	9,71	10,26	10,79	11,39	11,95	13,03	14,03	14,85	15,29		0,00	0,14	0,28	0,42	0,56	0,71	0,84	0,99	1,13	1,27
2300	5,89	6,40	6,89	7,44	8,10	8,74	9,35	9,94	10,49	11,02	11,61	12,17	13,21	14,15	14,85	15,08		0,00	0,15	0,29	0,44	0,59	0,74	0,88	1,03	1,18	1,33
2400	6,04	6,56	7,06	7,62	8,30	8,94	9,56	10,14	10,70	11,23	11,81	12,36	13,36	14,21	14,77	14,76		0,00	0,15	0,31	0,46	0,61	0,77	0,92	1,08	1,23	1,38
2500	6,17	6,70	7,22	7,79	8,48	9,13	9,75	10,34	10,89	11,41	11,99	12,52	13,47	14,23	14,62			0,00	0,16	0,32	0,48	0,64	0,80	0,96	1,12	1,28	1,44
2600	6,31	6,85	7,37	7,95	8,64	9,30	9,93	10,51	11,06	11,58	12,14	12,65	13,53	14,18	14,39			0,00	0,17	0,33	0,50	0,67	0,83	1,00	1,17	1,33	1,50
2700	6,43	6,98	7,51	8,10	8,80	9,46	10,09	10,67	11,21	11,72	12,26	12,74	13,56	14,09	14,09			0,00	0,17	0,35	0,52	0,69	0,87	1,04	1,21	1,39	1,56
2800	6,55	7,11	7,65	8,24	8,95	9,61	10,23	10,81	11,34	11,83	12,36	12,81	13,54	13,93				0,00	0,18	0,36	0,54	0,72	0,90	1,08	1,26	1,44	1,62
2900	6,66	7,23	7,77	8,37	9,08	9,74	10,36	10,93	11,45	11,93	12,42	12,85	13,48	13,71				0,00	0,19	0,37	0,56	0,74	0,93	1,11	1,30	1,49	1,67
3000	6,76	7,34	7,89	8,49	9,20	9,86	10,47	11,03	11,54	11,99	12,46	12,85	13,38	13,44				0,00	0,19	0,38	0,58	0,77	0,96	1,15	1,35	1,54	1,73
3100	6,86	7,44	8,00	8,60	9,31	9,97	10,57	11,12	11,61	12,04	12,47	12,82	13,23					0,00	0,20	0,40	0,60	0,79	1,00	1,19	1,39	1,59	1,79
3200	6,95	7,54	8,09	8,70	9,41	10,06	10,65	11,18	11,65	12,06	12,45	12,75	13,03					0,00	0,21	0,41	0,61	0,82	1,03	1,23	1,44	1,64	1,85
3300	7,04	7,62	8,18	8,79	9,49	10,13	10,71	11,23	11,67	12,05	12,40	12,65	12,79					0,00	0,21	0,42	0,63	0,84	1,06	1,27	1,48	1,69	1,90
3400	7,11	7,70	8,26	8,87	9,56	10,19	10,76	11,25	11,67	12,01	12,32	12,51	12,50					0,00	0,22	0,44	0,65	0,87	1,09	1,31	1,53	1,74	1,96
3500	7,18	7,77	8,33	8,93	9,62	10,24	10,79	11,25	11,64	11,95	12,21	12,34						0,00	0,22	0,45	0,67	0,90	1,12	1,34	1,57	1,80	2,02
3600	7,24	7,84	8,39	8,99	9,67	10,27	10,80	11,24	11,59	11,86	12,06	12,12						0,00	0,23	0,46	0,69	0,92	1,16	1,38	1,62	1,85	2,08

Belt drive vocabulary in four languages

English	German	French	Swedish
add-on power, additional kW rating	Übersetzungszuschlag m, Additionswert m pro Keilriemen	puissance f additionnelle	tilläggseffekt
additional kW rating, see add-on power			
adhesion rubber, see cushion rubber			
adjustable-speed belt, variator belt, variable speed belt	Variatorkeilriemen m	courroie f trapézoïdale pour variateurs de vitesse	variator(kil)rem, bredkilrem
agricultural V-belt, V-belt for agricultural machinery	Keilriemen m für die Land-wirtschaft, Keilriemen m für die Landtechnik	courroie f trapézoïdale pour machines agricoles	lantbrukskilrem, kilrem för lant-bruksmaskiner
angle of pulley groove	(Scheiben)rillenwinkel m	angle m d'une gorge de poulie	spårvinkel, profilvinkel (remskiva)
apparent coefficient of friction	scheinbare Reibungszahl f	coefficient m apparent de frottement	skenbart friktionstal
arc of contact angle, contact angle	Umschlingungswinkel m	angle m d'arc d'enroulement, arc m de contact	omfattningsvinkel, omslutningsvinkel
automotive V-belt	Keilriemen m für Kfz, Keilriemen m für den Kraftfahrzeugbau	courroie f trapézoïdale pour la construction automobile	fordonskilrem, kilrem för bilindustrin
axial force, axial load (variable speed pulley)	Scheibenspreizkraft f, Axialkraft f der Scheibe	force f axiale (poulie à vitesse variable)	axialkraft(variatorremskiva)
banded (V-)belt, see joined V-belt			
base length	Nennriemenlänge f	longueur f de base (nominale)	baslängd
base rubber, see bottom rubber			
basic power rating (per belt)	Nennleistung f (pro Riemen)	puissance f transmissible (par courroie)	baseffekt (per rem)

belt angle	Riemenwinkel *m*	angle *m* d'une courroie trapézoïdale	profilvinkel, remvinkel
belt dressing	Riemenappretur *f*, Riemenfett *n*	enduit *m* pour courroies	remsmörja, remvax
belt (pre-)tension, initial tension	Riemen(vor)spannung *f*	(pré-)tension *f* des courroies	rem(för)spänning, remkraft
belt tension ratio, see effective belt tension ratio			
bottom rubber, base rubber	Gummikern *m*, Füllgummi *n*	semelle *f* en caoutchouc	bottengummi, kärngummi
bushing pulley, bushing sheave, taper bored pulley	Scheibe *f* mit loser Nabe	poulie *f* à manchon, poulie *f* avec moyeu amovible	bussningsskiva
classical V-belt, conventional V-belt	Normalkeilriemen *m*, klassischer Keilriemen *m*	courroie *f* trapézoïdale classique	normalkilrem, klassisk kilrem
coefficient of friction, friction coefficient	Reibungszahl *f*	coefficient *m* de frottement	friktionstal
coefficient of traction	Durchzuggrad *m*	coefficient *m* de traction	belastningsgrad
cogged V-belt	gezahnter Keilriemen *m*	courroie *f* trapézoïdale à dents moulées	kuggad kilrem, kuggkilrem
contact angle, see arc of contact angle			
conventional V-belt, see classical V-belt			
conveyor V-belt	Förderkeilriemen *m*, Transportkeilriemen *m*	courroie *f* trapézoïdale transporteuse	transportkilrem
cord	Kord *m*	corde *f*	kord
crossed (belt) drive	verschränkter Antrieb *m* (180°)	courroie *f* croisée	helkorsad remdrift
crown	Bombierung *f*, Gewölbung *f*	bombage *m*	bombering
cushion rubber, adhesion rubber	Zwischengummi *n*, Haftmischung *f*	caoutchouc *m* intermédiaire	mellangummi
cut V-belt, raw edge V-belt, cut edge V-belt	flankenoffener Keilriemen *m*	courroie *f* trapézoïdale tranchée à flancs nus	skuren kilrem
datum circumference	Richtumfang *m*	circonférence *f* de référence	referensomkrets

datum diameter (pulley)	Richtdurchmesser m (Riemenscheibe)	diamètre m de référence (poulie à gorges)	referensdiameter (remskiva)
datum length	Richtlänge f	longueur f de référence	referenslängd
datum line differential	Richt-Linien-Differenz f	décalage m de la ligne de référence	referenslinjedifferens
datum width	Richtbreite f	largeur f de référence	referensbredd
design power	Berechnungsleistung f	puissance f effective, puissance f corrigée	drifteffekt
double (angle) V-belt, see hexagonal belt			
driven pulley	getriebene Scheibe f	poulie f menée, poulie f conduite	driven remskiva
driving pulley, driver	treibende Scheibe f	poulie f menante, poulie f conductrice	drivande remskiva
dual drive, see serpentine drive			
effective belt tension ratio, belt tension ratio	Trumkraftverhältnis n, Spannungsverhältnis n	rapport m de tension effectif par courroie	effektivkraftskvot
effective circumference	Bezugsumfang m	circonférence f effective	effektivomkrets
effective diameter (pulley)	Bezugsdurchmesser m (Keilriemenscheibe)	diamètre m effectif (poulie à gorges)	effektivdiameter (kilremsskiva)
effective length	Bezugslänge f	longueur f effective	effektivlängd, utvändig längd
effective line differential	Bezugs-Linien-Differenz f	décalage m de la ligne effective	effektivlinjedifferens
effective width	Bezugsbreite f	largeur f effective	effektivbredd
endless belt	endloser Riemen m	courroie f sans fin	ändlös rem
engine power (combustion engine)	Motorleistung f	puissance f nominale du moteur	motoreffekt, drivande maskinens effekt
fabric wrapping, fabric envelope, fabric jacket, textile wrapping	Umlagegewebe n, Hüllgewebe n, Deckgewebe n	renfort m supérieur	omläggsväv
fan belt	Ventilatorriemen m	courroie f de ventilateur	fläktrem
flat belt	Flachriemen m	courroie f plate	flat (driv)rem, flatrem

flat idler	zylindrische flache Spannrolle *f*, zylindrische flache Führungsrolle *f*	galet-tendeur *m* à jante plate	flat spännrulle, plan spännrulle
flat pulley	Flachscheibe *f*	poulie *f* pour courroies plates	flat remskiva, plan remskiva
friction coefficient, see coefficient of friction			
grooved pulley, see V-belt pulley			
height	Höhe *f*	hauteur *f*	remhöjd
hexagonal belt, double(angle) V-belt	Doppel(keil)riemen *m*, Hexagonalriemen *m*	courroie *f* hexagonale	hexagonalrem, dubbelkilrem, duplexrem, sexkantig rem
idler (pulley)	Spannrolle *f*, Führungsrolle *f*	galet-tendeur *m*, poulie *f* de tension	spännrulle
industrial V-belt	Industriekeilriemen *m*	courroie *f* trapézoïdale pour l'industrie	industrikilrem
initial tension, see belt (pre-)tension			
joined V-belt, multi-V-belt, banded (V-)belt	Verbundkeilriemen *m*, Kraftband *n*, Multi-V-Band *n*	courroie *f* trapézoïdale jumelée, courroie *f* trapézoïdale à brins multiples	samkilrem
link belt	Glieder-Keilriemen *m*	courroie *f* articulée	länkkilrem
matching number	Satznummer *f*, Differenznummer *f*	nombre *m* de différence	differensnummer
motor power (electric motor)	Motorleistung *f*	puissance *f* nominale du moteur	motoreffekt, drivande maskinens effekt
multiple V-belt drive	mehrrilliger Antrieb *m*, Mehrstrangantrieb *m*	transmission *f* à plusieurs courroies trapézoïdales	flerremsdrift, multipelremdrift
multi-V-belt, see joined V-belt			
narrow V-belt, wedge belt	Schmalkeilriemen *m*	courroie *f* trapézoïdale étroite	smalkilrem
neutral line, see pitch line			
neutral zone, see pitch zone			
number of teeth	Zähnezahl *f*	nombre *m* de dents	kuggtal, tandtal

open-end belt	endlicher Riemen m	courroie f ouverte	ändlig rem
pitch diameter (pulley)	Wirkdurchmesser m (Keilriemenscheibe)	diamètre m primitif (poulie à gorges)	delningsdiameter (kilremsskiva)
pitch length	Wirklänge f	longueur f primitive	delningslängd
pitch line, neutral line	Wirklinie f, neutrale Linie f	ligne f primitive	delningslinje, neutrallinje
pitch width	Wirkbreite f	largeur f primitive	delningsbredd
pitch zone, neutral zone	Wirkzone f	zone f primitive	delningsyta, delningsplan, neutral-zon
poly V-belt, see V-ribbed belt			
power rating, power transmission capability	Leistungswert m, Leistungsvermögen n	puissance f transmissible	effektöverföringsförmåga, effekttal
prime mover power	Übertragungsleistung f	puissance f nominale du moteur	drivande maskinens effekt
pulley, sheave (US)	Riemenscheibe f	poulie f	remskiva
quarter-turn (belt) drive	viertelverschränkter Antrieb m	transmission f quart détour, courroie f semi-croisée	halvkorsad remdrift
raw edge V-belt, see cut V-belt			
reinforcement	Armierung f, Verstärkung f	renforcement m	armering
relative height	relative Höhe f	hauteur f relative	relativ remhöjd
ride-out	Überstand m	dépassement m sur poulie	friliggning
serpentine drive, dual drive	Mehrscheibenantrieb m	transmission f serpentine	serpentindrift
service factor	Betriebsfaktor m, Belastungsfaktor m	facteur m de service	driftfaktor, maskinfaktor
shaft centre distance	Achsabstand m	entr'axe m	axelavstånd
sheave (US), see pulley			
side (tight and slack), strand	Trum m (straff/last und los/leer)	brin m (tendu et mou)	part (stram och slak), rempart
slack side	Leertrum m, Lostrum m	brin m mou	slak part
sliding slip	Gleitschlupf m	glissement m	glidslirning, totalslirning
slip	Schlupf m	glissement m	slirning
speed ratio	Übersetzung(sverhältnis n) f	rapport m de vitesse	utväxling
stepped pulley	Stufenscheibe f	poulie f etagée, poulie f à gradins	trappremskiva

strand, see side (tight and slack)			
synchronous belt, tooth(ed) belt, timing belt	Zahn(flach)riemen *m*, Synchronriemen *m*	courroie *f* synchrone, courroie *f* dentée	kuggrem, synkronrem
take-up allowance (shaft centre distance)	Verstellweg *m* (Achsabstand)	réglage *m* de l'entr'axe des poulies, course *f* de réglage (entr'axe)	ställbarhet (axelavstånd), justeringsmån (axelavstånd)
taper bored pulley, see bushing pulley			
tension slip	Dehnschlupf *m*	glissement *m* par tension	töjslirning
textile wrapping, see fabric wrapping			
tight side	Lasttrum *m*, Strafftrum *m*	brin *m* tendu	stram part, dragande part
timing belt, tooth(ed) belt, see synchronous belt			
top rubber	Gummiauflage *f*	caoutchouc *m* de dessus d'armature	toppgummi
top width	obere Breite *f*	largeur *f* au sommet	rembredd, toppbredd
twin V-belt	Zwillingkeilriemen *m*	courroie *f* trapézoïdale jumelée	tvillingkilrem, samkilrem
twisted belt drive	verschränkter Antrieb *m*	courroie *f* croisée	korsad remdrift
variable speed belt, variator belt, see adjustable-speed belt			
V-belt	Keilriemen *m*	courroie *f* trapézoïdale	kilrem
V-belt drive	Keilriemenantrieb *m*	transmission *f* par courroie trapézoïdale	kilremsdrift, kilremsväxel
V-belt for agricultural machinery, see agricultural V-belt			
V-ribbed belt, poly-V-belt	Poly-V-Riemen *m*, Rippenband *n*	courroie *f* striée, courroie *f* poly-V	skårrem, ribbrem, poly-V-rem
V-belt pulley, grooved pulley, wedge belt pulley	Keil(riemen)scheibe *f*	poulie *f* à gorges	kilremsskiva, spårad remskiva
wedge belt, see narrow V-belt			
wedge belt pulley, see V-belt pulley			
wide V-belt	Breitkeilriemen *m*	courroie *f* trapézoïdale large	bredkilrem, variator(kil)rem

Subject index

286